Measure Theory

Measure Theory
A First Course

Author
Carlos S. Kubrusly

ELSEVIER

AMSTERDAM • BOSTON • HEIDELBERG • LONDON
NEW YORK • OXFORD • PARIS • SAN DIEGO
SAN FRANCISCO • SINGAPORE • SYDNEY • TOKYO

Academic Press is an imprint of Elsevier

Elsevier Academic Press

30 Corporate Drive, Suite 400, Burlington, MA 01803, USA

525 B Street, Suite 1900, San Diego, California 92101-4495, USA

84 Theobald's Road, London WC1X 8RR, UK

This book is printed on acid-free paper.

Library of Congress Cataloging-in-Publication Data
APPLICATION SUBMITTED

British Library Cataloguing in Publication Data
A catalogue record for this book is available from the British Library

ISBN 13: 978-0-12-370899-1
ISBN 10: 0-12-370899-0

For all information on all Elsevier Academic Press publications
visit our Web site at www.books.elsevier.com

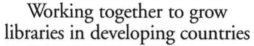

To You, then and now

Preface

This is a textbook for a first course in measure theory with integration. It offers an abstract approach for an introduction to measure and integration in nine chapters, where the classical cases of Lebesgue measure and Lebesgue integral are also presented as a specially important particular case of the general theory. Chapter 1 introduces the reader to σ-algebras and measurable functions, Chapter 2 considers the concept of measure, and the integral of nonnegative measurable functions with respect to a given measure is focused on in Chapter 3. This is extended in Chapter 4 to real-valued (not necessarily nonnegative) measurable functions, and L^p spaces are studied in Chapter 5. Convergence concepts for sequences of measurable functions are examined and compared in Chapter 6, and classic decompositions of measures are investigated in Chapter 7. Extension theorems make the subject of Chapter 8, where Lebesgue measure is constructed and discussed in detail. Finally, product measures and integrals with respect to them close the book in Chapter 9.

The last section of each chapter is a section of Problems. These are not just routine exercises but are an integral part of the main text, consisting of auxiliary results, extensions of the theory, examples, and counterexamples. The better half of the problems are accompanied by hints, and the reader is urged to follow these sections with the same care expected for a conventional theory section. In fact, part of the theory is sometimes shifted to the Problems section. When this happens, all problems are filled with hints in order to persuade readers to take an active part in theory construction, offering them a real chance to prove statements that are new to them, thus acquiring a firmer grasp of the theory they helped to build.

At the end of each chapter the reader will find a collection of suggested readings. This comprises books only (no research papers are mentioned at this introductory level) and has a double purpose: to indicate where different approaches and proofs can be found, and also to point out the way to further and more general results. In this sense, some of the references are suggested as a second reading on the subject.

The material in this book was prepared to be covered in a one-semester beginning graduate course. The resulting text is the outcome of attempts to meet the needs of a contemporary first course in measure theory for mathematicians that is also accessible to a wider audience composed of mathematics, statistics, economics, engineering, and physics students. I tried to respond to the input from those students by offering a text that contains complete proofs (sketchy proofs seem to have been a perpetual complaint) and a collection of problems that includes several questions asked by them throughout the years. Although naturally addressed to graduate students, it is certainly accessible to advanced undergraduate students too. Actually, the book is rather self-contained, and the prerequisites are very modest, namely, the usual undergraduate introductory analysis and, just for Chapter 5, linear spaces as it is usually taught in standard linear algebra courses. No acquaintance with functional analysis is required but, of course, elementary set theory is necessary. For instance, I assume the reader is familiar with the notion of cardinality and with the concept of countable and uncountable sets, knows that the set of all rational numbers is countable, and will quickly recall the following basic facts: a countable union of countable sets is countable, every infinite set (in particular, every uncountable set) has a countably infinite proper subset, and every countably infinite family can be reenumerated into a sequence.

I have been lecturing on this subject for a long time, so I benefited from the help of many friends, among students and colleagues, and I am really grateful to all of them; in particular to Renato A.A. da Costa, Leonardo B. Gonçalves, and Alexandre Street, who helped with the quest for typos. Special thanks are due to Jessica Q. Kubrusly, from whom I stole the notes for Section 7.3, and to Adrian H. Pizzinga, who read the entire text, corected a number of typos and inaccuracies, and remained good-natured when I accepted just almost all of his many suggestions. Thanks are also due to my friend and colleague Marcelo D. Fragoso, who was indeed an accomplice in writing this book. I am grateful to Catholic University of Rio de Janeiro for providing the release time that made this project possible, as well as to CNPq (Brazilian National Research Council) for a research grant.

Carlos S. Kubrusly
Rio de Janeiro
July 2006

Contents

1
Measurability

1.1 σ-Algebra

Starting with Chapter 2, we shall be dealing with functions of sets (i.e., with functions whose domains are sets of sets). In principle, we might argue that a candidate for the domain of our functions of sets would be the *power set* $\wp(X)$ of a given set X (i.e., the collection of all subsets of X). This indeed is a possible candidate, but in some instances the power set is too large a set for the domain of our functions. What does this mean? It means that some functions may lose essential properties if their domain is too large or, in other words, some of the functions we wish to consider would not be "well-behaved" when defined on a domain that is too big (we discuss this in Chapter 8). Given a set X, a collection of subsets of X (i.e., a subcollection of the power set $\wp(X)$) that will fit our needs is called a σ-algebra.

Definition 1.1. A collection $\mathcal{A}$ of subsets of a set X is an *algebra* (or a *field*, or a *Boolean algebra*) of sets if it satisfies the following axioms.

  (a) The whole set X and the empty set ∅ belong to $\mathcal{A}$.

  (b) The complement $X \backslash E$ of a set E in $\mathcal{A}$ belongs to $\mathcal{A}$.

  (c) The union of a finite collection of sets in $\mathcal{A}$ belongs to $\mathcal{A}$.

A *σ-algebra* (or a *σ-field*, or still a *Borel field*) is an algebra $\mathcal{X}$ of subsets of a set X for which axiom (c) is extended as follows.

  (c′) The union of a countable collection of sets in $\mathcal{X}$ belongs to $\mathcal{X}$.

By a *measurable space* we mean a pair $(X, \mathcal{X})$ consisting of an arbitrary set X and a σ-algebra $\mathcal{X}$ of subsets of X. Sets in $\mathcal{X}$ are called *measurable sets* (with respect to the σ-algebra $\mathcal{X}$) or *$\mathcal{X}$-measurable* sets.

Axiom (b) and the De Morgan laws (viz., $X \backslash (\bigcup_\alpha A_\alpha) = \bigcap_\alpha (X \backslash A_\alpha)$ and $X \backslash (\bigcap_\alpha A_\alpha) = \bigcup_\alpha (X \backslash A_\alpha)$ for any collection $\{A_\alpha\}$ of subsets of a set X) ensure that in Definition 1.1 (of an algebra and a σ-algebra) axioms (c) and (c$'$) can be replaced, respectively, by the following equivalent axioms.

(c) The intersection of a finite collection of sets in $\mathcal{A}$ belongs to $\mathcal{A}$.

(c$'$) The intersection of a countable collection of sets in $\mathcal{X}$ belongs to $\mathcal{X}$.

Example 1A. Let X be an arbitrary set. The power set $\wp(X)$ is clearly a σ-algebra of subsets of X. Indeed, it is more than that: the union of an arbitrary collection (not necessarily countable) of sets in $\wp(X)$ belongs to $\wp(X)$. In fact, the power set $\wp(X)$ is the largest σ-algebra of subsets of X, and the collection $\{\varnothing, X\}$ is the smallest (in the inclusion ordering). Recall that a *partition* of a set X is any collection of pairwise disjoint subsets of X that cover X. If $\{A, B\}$ is a partition of X (so that $A \cup B = X$ and $A \cap B = \varnothing$), then $\mathcal{X} = \{\varnothing, A, B, X\}$ also is a σ-algebra of subsets of X.

It is readily verified that *the intersection of any collection of σ-algebras of subsets of a set X is again a σ-algebra of subsets of X*. Given a nonempty collection $\mathcal{C}$ of subsets of a set X (i.e., $\mathcal{C} \subseteq \wp(X)$), there exists a smallest (inclusion ordering) σ-algebra $\mathcal{X}_\mathcal{C}$ of subsets of X that includes $\mathcal{C}$. That is, $\mathcal{X}_\mathcal{C}$ is included in any σ-algebra of subsets of X that includes $\mathcal{C}$ (if $\mathcal{X}$ is a σ-algebra of subsets of X and $\mathcal{C} \subseteq \mathcal{X}$, then $\mathcal{X}_\mathcal{C} \subseteq \mathcal{X}$). Indeed, let $\mathbf{X}_\mathcal{C}$ denote the collection of all σ-algebras of subsets of X that include $\mathcal{C}$ (which is non-empty because the power set $\wp(X)$ clearly is an element of $\mathbf{X}_\mathcal{C}$). Now put $\mathcal{X}_\mathcal{C} = \bigcap \mathbf{X}_\mathcal{C}$, the intersection of all σ-algebras of subsets of X that include $\mathcal{C}$. This smallest σ-algebra $\mathcal{X}_\mathcal{C}$ is called *the σ-algebra generated by $\mathcal{C}$*.

An especially important particular case is the σ-algebra $\Re$ generated by the collection of all open intervals of the real line $\mathbb{R}$ (or of all closed intervals, what boils down to the same thing — why?) is called the *Borel algebra*. The elements of $\Re$ (i.e., the $\Re$-measurable sets) are called *Borel sets*.

Remark: Since $A \backslash B = A \cap X \backslash B$ for every $A, B \subseteq X$, axioms (a), (b), and (c) are equivalent to axioms (a$'$), (b$'$), and (c), where (a$'$) and (b$'$) are given as follows.

(a$'$) X belongs to $\mathcal{A}$.

(b$'$) The difference $E \backslash F$ of sets E and F in $\mathcal{A}$ belongs to $\mathcal{A}$.

A nonempty collection of sets that satisfies axioms (b') and (c) is called a *ring* (or a *Boolean ring*) of sets. If it satisfies axioms (b') and (c'), then it is a *σ-ring*. Every algebra (every σ-algebra) is a ring (a σ-ring). If a ring (a σ-ring) of subsets of a set X contains X, then it is an algebra (a σ-algebra).

1.2 Real-Valued Functions

Definition 1.2. Let $(X, \mathcal{X})$ be a measurable space. A function $f \colon X \to \mathbb{R}$ is *measurable* (with respect to the σ-algebra $\mathcal{X}$), or $\mathcal{X}$-*measurable*, if the inverse image of (α, ∞) under f is a measurable set for any real number α:

$$f^{-1}\big((\alpha, \infty)\big) = \{x \in X \colon f(x) > \alpha\} \in \mathcal{X} \quad \text{for every} \quad \alpha \in \mathbb{R}.$$

Remark: The sign $>$ in Definition 1.2 can be replaced with $\geq$, $<$, or $\leq$, yielding equivalent definitions of a measurable function (Problem 1.1).

Example 1B. Let $\{X, \mathcal{X}\}$ be a measurable space and let E be a measurable set (i.e., take any $E \in \mathcal{X}$). Consider the *characteristic function* of E; that is, consider the map $\chi_E \colon X \to \{0, 1\}$ such that

$$\chi_E(x) = \begin{cases} 1, & x \in E, \\ 0, & x \in X \backslash E. \end{cases}$$

It is readily verified that $\chi_E \colon X \to \{0, 1\}$ is a measurable function. Indeed,

$$\chi_E^{-1}\big((\alpha, \infty)\big) = \{x \in X \colon \chi_E(x) > \alpha\} = \begin{cases} \varnothing, & 1 \leq \alpha, \\ E, & 0 \leq \alpha < 1, \\ X, & \alpha < 0. \end{cases}$$

In particular, *every constant function is measurable*. In fact, the function $1 \colon X \to \mathbb{R}$ such that $1(x) = 1$ for all $x \in X$ coincides with the characteristic function χ_X. Moreover, it is readily verified that γf is measurable for every $\gamma \in \mathbb{R}$ whenever $f \colon X \to \mathbb{R}$ is a measurable function. This proves the italicized result once a constant function is precisely $\gamma \chi_X$ for some $\gamma \in \mathbb{R}$.

Example 1C. Let $\mathbb{N}$ be the set of all positive integers, and let $\mathbb{N}_e$ and $\mathbb{N}_o$ denote the subsets of $\mathbb{N}$ consisting of all even and odd numbers, respectively. Since $\{\mathbb{N}_o, \mathbb{N}_e\}$ forms a partition of $\mathbb{N}$, it follows by Example 1A that $\mathcal{N} = \{\varnothing, \mathbb{N}_o, \mathbb{N}_e, \mathbb{N}\}$ is a σ-algebra of subsets of $\mathbb{N}$. Consider the measurable space $\{\mathbb{N}, \mathcal{N}\}$ and the function $f \colon \mathbb{N} \to \mathbb{N} \subset \mathbb{R}$ defined by $f(n) = n$ for every $n \in \mathbb{N}$. Note that f is not measurable (i.e., not $\mathcal{N}$-measurable). Indeed,

$$f^{-1}\big((1, \infty)\big) = \{n \in \mathbb{N} \colon f(n) > 1\} = \{2, 3, 4, \ldots\} \notin \mathcal{N}.$$

Let $\mathcal{X}$ be a σ-algebra of subsets of a set X, take an arbitrary measurable set E of $\mathcal{X}$ (i.e., take any $E \in \mathcal{X}$), and put

$$\mathcal{E} = \wp(E) \cap \mathcal{X},$$

where $\wp(E)$ is the power set of E. This is the collection of all $\mathcal{X}$-measurable subsets of E, which is a σ-algebra of subsets of E. Indeed, since $\mathcal{X}$ is a σ-algebra, it follows at once that a countable union of sets in $\mathcal{E}$ belongs to $\mathcal{E}$. Moreover, if $B \in \mathcal{E}$, then $B \subseteq E$ and $B \in \mathcal{X}$, and hence $E \backslash B \in \wp(E)$ and $E \backslash B = E \cap (X \backslash B) = X \backslash [(X \backslash E) \cup B] \in \mathcal{X}$. Note that the smallest σ-algebra of subsets of E, namely $\{E, \varnothing\}$, is a subcollection of $\mathcal{E}$.

Proposition 1.3. *The restriction of an $\mathcal{X}$-measurable function to an $\mathcal{X}$-measurable set E is an $\mathcal{E}$-measurable function.*

Proof. Let $\mathcal{X}$ be a σ-algebra of subsets of a set X, take an $\mathcal{X}$-measurable function $f : X \to \mathbb{R}$, let E be an $\mathcal{X}$-measurable set, and consider the restriction $f|_E : E \to \mathbb{R}$ of f to E (i.e., $f|_E(e) = f(e)$ for every $e \in E$). Take an arbitrary $\alpha \in \mathbb{R}$ and observe that

$$\{e \in E\colon f|_E(e) > \alpha\} = \{x \in X\colon f(x) > \alpha\} \cap E \quad \text{lies in} \quad \mathcal{E} = \wp(E) \cap \mathcal{X}.$$

Indeed, as a subset of E, this obviously lies in $\wp(E)$, and it lies in $\mathcal{X}$ because it is the intersection of two $\mathcal{X}$-measurable sets. Therefore, $f|_E$ is an $\mathcal{E}$-measurable function. $\qquad\square$

Proposition 1.4. *Let $\{X, \mathcal{X}\}$ be a measurable space. Suppose A and B are $\mathcal{X}$-measurable sets and consider the σ-algebras $\mathcal{A} = \wp(A) \cap \mathcal{X}$ and $\mathcal{B} = \wp(B) \cap \mathcal{X}$ of subsets of A and B, respectively. Take a function $f : X \to \mathbb{R}$ and consider its restrictions $f|_A : A \to \mathbb{R}$ and $f|_B : B \to \mathbb{R}$ to A and B, respectively. If $A \cup B = X$, then f is $\mathcal{X}$-measurable if and only if $f|_A$ is $\mathcal{A}$-measurable and $f|_B$ is $\mathcal{B}$-measurable.*

Proof. Proposition 1.3 ensures the "only if" part. To prove the "if" part, take an arbitrary $\alpha \in \mathbb{R}$ and observe that, since $A \cup B = X$,

$$\{x \in X\colon f(x) > \alpha\} = \{a \in A\colon f|_A(a) > \alpha\} \cup \{b \in B\colon f|_B(b) > \alpha\},$$

which is the union of an $\mathcal{A}$-measurable set and a $\mathcal{B}$-measurable set (reason: $f|_A$ is an $\mathcal{A}$-measurable function and $f|_B$ is an $\mathcal{B}$-measurable function). Since $\mathcal{A}$ and $\mathcal{B}$ are included in $\mathcal{X}$, it follows that both sets belong to $\mathcal{X}$, and hence their union $\{x \in X\colon f(x) > \alpha\}$ is an $\mathcal{X}$-measurable set, which means that f is an $\mathcal{X}$-measurable function. $\qquad\square$

There exists an abundant supply of measurable functions (e.g., see Problem 1.2). Actually, let $\mathcal{X}$ be a σ-algebra of subsets of a (nonempty) set X and take a couple of measurable functions, say $f \colon X \to \mathbb{R}$ and $g \colon X \to \mathbb{R}$, with respect to $\mathcal{X}$. Recall that a polynomial $p(f,g)$ of f and g (with real coefficients) is any (finite) linear combination of $\{f^n g^m\}_{n,m \geq 0}$.

Proposition 1.5. *If f and g are measurable, then so is every $p(f,g)$.*

Proof. Let f and g be measurable functions and let γ be any real number. We have already verified that γ (a constant function) and γf (a multiple of a measurable function) are measurable functions. Thus it is enough to show that $f + g$ and fg are measurable functions in order to ensure that all finite words $p(f,g)$ (i.e., all linear combinations of $\{f^n g^m\}_{n,m \geq 0}$) are measurable functions. Let $\mathbb{Q}$ denote the rational field, take an arbitrary $\rho \in \mathbb{Q}$, an arbitrary $\alpha \in \mathbb{R}$, and note that $f^{-1}((\rho, \infty)) \cap g^{-1}((\alpha - \rho, \infty))$ is a measurable set (it is the intersection of measurable sets because f and g are measurable functions). Denote this measurable set by $E_{\alpha, \rho}$ so that

$$E_{\alpha, \rho} = \big\{x \in X \colon f(x) > \rho \ \text{and} \ g(x) > \alpha - \rho\big\}.$$

For each $\alpha \in \mathbb{R}$, observe that $(f + g)(x) = f(x) + g(x) > \alpha$ if and only if $f(x) > \rho$ and $g(x) > \alpha - \rho$ for some $\rho \in \mathbb{Q}$. That is, if and only if $x \in E_{\alpha, \rho}$ for some $\rho \in \mathbb{Q}$. Therefore, for any $\alpha \in \mathbb{R}$,

$$(f + g)^{-1}((\alpha, \infty)) = \{x \in X \colon (f + g)(x) > \alpha\} = \bigcup_{\rho} E_{\alpha, \rho}.$$

Since each $E_{\alpha, \rho}$ is a measurable set and since $\mathbb{Q}$ is a countable set, it follows that $\bigcup_{\rho \in \mathbb{Q}} E_{\alpha, \rho}$ is a measurable set for every $\alpha \in \mathbb{R}$, and hence $f + g$ is a measurable function (Definitions 1.1 and 1.2). Now observe that f^2 (defined by $f^2(x) = f(x)^2$ for every $x \in X$) also is a measurable function. Indeed, $\{x \in X \colon f^2(x) > \alpha\} = X$ if $\alpha < 0$ and $f(x)^2 > \alpha$ if and only if either $f(x) > \sqrt{\alpha}$ or $f(x) < -\sqrt{\alpha}$ whenever $\alpha \geq 0$, which implies that the set $(f^2)^{-1}((\alpha, \infty)) = \{x \in X \colon f^2(x) > \alpha\}$ is measurable:

$$(f^2)^{-1}((\alpha, \infty)) = \begin{cases} X, & \alpha < 0, \\ f^{-1}((\sqrt{\alpha}, \infty)) \cup (-f)^{-1}((\sqrt{\alpha}, \infty)), & \alpha \geq 0, \end{cases}$$

(union of measurable sets). Thus f^2 is a measurable function. But

$$fg = \tfrac{1}{4}\big((f + g)^2 - (f - g)^2\big),$$

which involves multiplication by a constant, addition, and squaring of measurable functions, and therefore fg is a measurable function. $\square$

Let $X^{\mathbb{R}}$ denote the real linear space of all real-valued functions on a set X. If $\mathcal{X}$ is a σ-algebra of subsets of X, then Proposition 1.5 ensures that the collection of all $\mathcal{X}$-measurable functions forms a linear subspace of $X^{\mathbb{R}}$, thus a linear space itself. Now take an arbitrary function $f: X \to \mathbb{R}$ and set

$$F^+ = \{x \in X : f(x) \geq 0\} \quad \text{and} \quad F^- = \{x \in X : f(x) \leq 0\}.$$

Let $\mathcal{X}_{F^+}$ and $\mathcal{X}_{F^-}$ be the characteristic functions of F^+ and F^-, respectively, and consider the real-valued functions on X,

$$f^+ = f\chi_{F^+} \quad \text{and} \quad f^- = -f\chi_{F^-},$$

which are called the *positive part* of f and the *negative part* of f, respectively. Note that f^+ and f^- are both nonnegative functions (i.e., $f^+(x) \geq 0$ and $f^-(x) \geq 0$ for every $x \in X$). Moreover, the functions f and its *absolute value* $|f|$ (given by $|f|(x) = |f(x)|$ for every $x \in X$) can be expressed in terms of the positive and negative parts of f as follows:

$$f = f^+ - f^- \quad \text{and} \quad |f| = f^+ + f^-.$$

Note that these can be reversed to yield

$$f^+ = \tfrac{1}{2}(|f| + f) \quad \text{and} \quad f^- = \tfrac{1}{2}(|f| - f).$$

Proposition 1.6. *The following assertions are pairwise equivalent.*

(a) *f is a measurable function.*

(b) *f^+ and f^- are measurable functions.*

(c) *F^+ and F^- are measurable sets and $|f|$ is a measurable function.*

Proof. Suppose f is a measurable function (with respect to a σ-algebra $\mathcal{X}$ of subsets of X). In this case, F^+ and F^- are both measurable sets (i.e., they belong to $\mathcal{X}$ — Definition 1.2), and therefore χ_{F^+} and χ_{F^-} are measurable functions (with respect to the same σ-algebra $\mathcal{X}$ — Example 1B). Thus Proposition 1.5 ensures that $f^+ = f\chi_{F^+}$ and $f^- = -f\chi_{F^-}$ are measurable functions, and so is $|f| = f^+ + f^-$. Note that the relation $f = f^+ - f^-$ is enough to guarantee the converse: f is measurable whenever f^+ and f^- are measurable functions. We proved so far that (a) and (b) are equivalent, and (a) implies (c). Now we shall verify that (c) implies (a). First observe that the restrictions of $|f|$ to F^+ and F^- coincide with the restrictions of f to F^+ and with the restriction of $-f$ to F^-, respectively; that is,

$$f|_{F^+} = |f|\big|_{F^+} : F^+ \to \mathbb{R} \quad \text{and} \quad f|_{F^-} = -|f|\big|_{F^-} : F^- \to \mathbb{R}.$$

Since $F^+ \cup F^- = X$, it follows by Proposition 1.4 that, if F^+ and F^- are $\mathcal{X}$-measurable sets and $|f|$ is an $\mathcal{X}$-measurable function, then $f|_{F^+}$ is $\mathcal{F}^+$-measurable and $-f|_{F^-}$ (and so $f|_{F^-}$) is $\mathcal{F}^-$-measurable, with the σ-algebras $\mathcal{F}^+$ and $\mathcal{F}^-$ given by $\wp(F^+) \cap \mathcal{X}$ and $\wp(F^-) \cap \mathcal{X}$, respectively. Another application of Proposition 1.4 ensures that f is $\mathcal{X}$-measurable. $\square$

It is worth noticing that "$|f|$ measurable" does not imply "f measurable" (i.e., (c) does not imply (a) without the assumption that F^+ and F^- are measurable). For instance, take a measurable space $(X, \mathcal{X})$ and suppose there exists a partition $\{A, A'\}$ of X, where A and A' are not $\mathcal{X}$-measurable sets, and consider the function $f \colon X \to \mathbb{R}$ given by $f(x) = 1$ for $x \in A$ and $f(x) = -1$ for $x \in A'$ so that $F^+ = A$ and $F^- = A'$. It is plain that f is not an $\mathcal{X}$-measurable function (for A is not an $\mathcal{X}$-measurable set). However, $|f|$ is a constant function, thus measurable (with respect to any σ-algebra).

1.3 Extended Real-Valued Functions

We shall be dealing with unbounded subsets of $\mathbb{R}$. In particular, we shall be dealing with inf and sup of unbounded subsets of $\mathbb{R}$ as well as with the notion of length of unbounded sets such as $\mathbb{R}$ itself. This lead us to introduce the *extended real number system* (or the *extended real line*), which is the collection $\overline{\mathbb{R}} = \mathbb{R} \cup \{-\infty, +\infty\}$ consisting of the real field $\mathbb{R}$ and a pair of symbols, namely, $-\infty$ and $+\infty$. These symbols are not numbers, certainly not real numbers. Actually, we might adopt any other pair of symbols, such as, for instance, (α, ω) or $(\triangleleft, \triangleright)$ instead of the "set-down-eights" $(-\infty, +\infty)$, but we will stick with the usual notation. Regarding the natural ordering of the real line $\mathbb{R}$, we declare that $-\infty < x < +\infty$ for all x in $\mathbb{R}$, thus extending the natural order of $\mathbb{R}$ to $\overline{\mathbb{R}}$. Note that $\overline{\mathbb{R}}$ is not a field, even though arithmetics with the new symbols are partially defined in the usual fashion, with some exceptions (e.g., $-\infty$ and $+\infty$ cannot be added together in any order; equivalently, the subtraction of $+\infty$ with itself is not defined).

The definition of measurable function for extended real-valued functions is exactly the same as in Definition 1.2: if $(X, \mathcal{X})$ is a measurable space, then an extended real-valued function $f \colon X \to \overline{\mathbb{R}}$ is *measurable* (with respect to the σ-algebra $\mathcal{X}$), or $\mathcal{X}$-*measurable*, if the inverse image of (α, ∞) under f is a measurable set for any real number (i.e., for any $\alpha \in \mathbb{R}$):

$$f^{-1}\big((\alpha, \infty)\big) = \big\{x \in X \colon f(x) > \alpha\big\} \in \mathcal{X} \quad \text{for every} \quad \alpha \in \mathbb{R}.$$

As before, the sign $>$ in the foregoing expression can be replaced with $\geq$, $<$, or $\leq$, yielding equivalent definitions of an extended real-valued measurable function (Problem 1.1). If f is an extended real-valued $\mathcal{X}$-measurable function, then it is easy to verify (Problem 1.4) that

$$F_{+\infty} = \big\{x \in X \colon f(x) = +\infty\big\} \quad \text{and} \quad F_{-\infty} = \big\{x \in X \colon f(x) = -\infty\big\}$$

are $\mathcal{X}$-measurable sets. Let $\chi_{X\setminus(F_{+\infty}\cup F_{-\infty})}$ be the characteristic function of the complement of $F_{+\infty} \cup F_{-\infty}$ and consider the real-valued function

$$f_{\mathbb{R}} = f\chi_{X\setminus(F_{+\infty}\cup F_{-\infty})} \colon X \to \mathbb{R}$$

(i.e., $f_{\mathbb{R}}(x) = f(x)$ if $x \notin (F_{+\infty} \cup F_{-\infty})$ and $f_{\mathbb{R}}(x) = 0$ if $x \in (F_{+\infty} \cup F_{-\infty})$; in other words, $f_{\mathbb{R}}(x) = f(x)$ if $f(x) \neq \pm\infty$, and $f_{\mathbb{R}}(x) = 0$ if $f(x) = \pm\infty$).

Proposition 1.7. *An extended real-valued function $f \colon X \to \overline{\mathbb{R}}$ is $\mathcal{X}$-measurable if and only if the real-valued function $f_{\mathbb{R}} \colon X \to \mathbb{R}$ is $\mathcal{X}$-measurable and both sets $F_{+\infty}$ and $F_{-\infty}$ are $\mathcal{X}$-measurable.*

Proof. Take an arbitrary $\alpha \in \mathbb{R}$ and note that

$$\big\{x \in X \colon f(x) > \alpha\big\} = \begin{cases} \big\{x \in X \colon f_{\mathbb{R}}(x) > \alpha\big\} \cup F_{+\infty}, & \alpha \geq 0, \\ \big\{x \in X \colon f_{\mathbb{R}}(x) > \alpha\big\} \setminus F_{-\infty}, & \alpha < 0. \end{cases}$$

Thus f is a measurable function whenever $f_{\mathbb{R}}$ is a measurable function and $F_{+\infty}$ and $F_{-\infty}$ are measurable sets (recall: $A\setminus B = X\setminus[(X\setminus A)\cup B]$). Since

$$\big\{x \in X \colon f_{\mathbb{R}}(x) > \alpha\big\} = \begin{cases} \big\{x \in X \colon f(x) > \alpha\big\} \setminus F_{+\infty}, & \alpha \geq 0, \\ \big\{x \in X \colon f(x) > \alpha\big\} \cup F_{-\infty}, & \alpha < 0, \end{cases}$$

and since $F_{+\infty}$ and $F_{-\infty}$ are measurable sets whenever f is a measurable function, it follows that $f_{\mathbb{R}}$ is measurable whenever f is. $\qquad\square$

If S is any set, then an *S-valued sequence* (or a sequence of elements in S) is just an S-valued function defined on $\mathbb{N}$ (the natural numbers, or on $\mathbb{N}_0 = \{0\} \cup \mathbb{N}$). Consider an $\overline{\mathbb{R}}$-valued sequence $\{\alpha_n\}$. As usual, let $\inf_n \alpha_n$ and $\sup_n \alpha_n$ be the greatest lower bound and the least upper bound of $\{\alpha_n\}$, respectively, which exist and are unique in $\overline{\mathbb{R}}$. The *limit inferior* and *limit superior* of $\{\alpha_n\}$ are defined in $\overline{\mathbb{R}}$, respectively, by

$$\liminf_n \alpha_n = \sup_n \inf_{n \leq k} \alpha_k \quad \text{and} \quad \limsup_n \alpha_n = \inf_n \sup_{n \leq k} \alpha_k.$$

If $\liminf_n \alpha_n = \limsup_n \alpha_n = \alpha$, then we say that $\{\alpha_n\}$ *converges* to $\alpha \in \overline{\mathbb{R}}$ and write $\lim_n \alpha_n = \alpha$. For an $\mathbb{R}$-valued sequence this is equivalent to the standard definition of convergence. That is, a real-valued sequence $\{\alpha_n\}$ converges to $\alpha \in \mathbb{R}$ if for every $\varepsilon > 0$ there exists an integer $n_\varepsilon \geq 1$ such that

$|\alpha_n - \alpha| < \varepsilon$ whenever $n \geq n_\varepsilon$. Moreover, if $\{\alpha_n\}$ is a real-valued *bounded* sequence (i.e., if each α_n lies in $\mathbb{R}$ and $\sup_n |\alpha_n|$ also lies in $\mathbb{R}$), then the sequences $\{\inf_{n \leq k} \alpha_k\}$ and $\{\sup_{n \leq k} \alpha_k\}$ converge in $\mathbb{R}$ to $\liminf_n \alpha_n$ and $\limsup_n \alpha_n$, respectively (which are both real numbers). In this case,

$$\liminf_n \alpha_n = \lim_n \inf_{n \leq k} \alpha_k \quad \text{and} \quad \limsup_n \alpha_n = \lim_n \sup_{n \leq k} \alpha_k.$$

Now take $f_n\colon X \to \overline{\mathbb{R}}$ for each $n \in \mathbb{N}$ so that $\{f_n\}$ is a sequence of extended real-valued functions on X. Since each $f_n(x)$ lies in $\overline{\mathbb{R}}$ for every x in X, put

$$\phi(x) = \inf_n f_n(x), \qquad \Phi(x) = \sup_n f_n(x),$$

$$\underline{f}(x) = \liminf_n f_n(x), \qquad \overline{f}(x) = \limsup_n f_n(x),$$

for every $x \in X$, which always exist as elements of $\overline{\mathbb{R}}$, and therefore they define the functions $\phi\colon X \to \overline{\mathbb{R}}$, $\Phi\colon X \to \overline{\mathbb{R}}$, $\underline{f}\colon X \to \overline{\mathbb{R}}$, and $\overline{f}\colon X \to \overline{\mathbb{R}}$. If $\underline{f}(x) = \overline{f}(x)$ for every $x \in X$, then the $\overline{\mathbb{R}}$-valued sequence $\{f_n(x)\}$ converges in $\overline{\mathbb{R}}$ for every $x \in X$. The common quantity $\underline{f}(x) = \overline{f}(x) \in \overline{\mathbb{R}}$, denoted by $f(x)$ for each $x \in X$, is called the *limit* of the $\overline{\mathbb{R}}$-valued sequence $\{f_n(x)\}$. This establishes a function $f\colon X \to \overline{\mathbb{R}}$. In this case (i.e., when $\underline{f} = \overline{f}$), we say that the sequence of functions $\{f_n\}$ *converges pointwise* to the *limit function* f and write $f = \lim_n f_n$, which means that, for every $x \in X$,

$$f(x) = \lim_n f_n(x).$$

Again, if $\{f_n\}$ is a sequence of $\mathbb{R}$-valued functions, then this coincides with the standard definition of pointwise convergence: a sequence of real-valued functions $f_n\colon X \to \mathbb{R}$ on X converges pointwise to a function $f\colon X \to \mathbb{R}$ if $|f_n(x) - f(x)| \to 0$ as $n \to \infty$ for every $x \in X$ (i.e., for each $\varepsilon > 0$ and each $x \in X$ there is an $n_{\varepsilon,x} \in \mathbb{N}$ such that $|f_n(x) - f(x)| \leq \varepsilon$ whenever $n \geq n_{\varepsilon,x}$).

Proposition 1.8. *If every f_n is measurable, then so are ϕ, Φ, $\underline{f}$ and $\overline{f}$. Moreover, if $f = \lim_n f_n$, then f is measurable.*

Proof. Recall that an arbitrary intersection of closed sets is closed and that an arbitrary union of open sets is open. Take any $\alpha \in \mathbb{R}$, and note that

$$\{x \in X\colon \phi(x) \geq \alpha\} = \bigcap_n \{x \in X\colon f_n(x) \geq \alpha\},$$

$$\{x \in X\colon \Phi(x) > \alpha\} = \bigcup_n \{x \in X\colon f_n(x) > \alpha\}.$$

Since $\{f_n\}$ is a sequence of measurable functions, the preceding sets are measurable (reason: they are made up of a countable intersection and a countable union of measurable sets, respectively, thus measurable themselves), and hence ϕ and Φ are measurable functions. Now put

$$\phi_n(x) = \inf_{n \leq k} f_k(x) \quad \text{and} \quad \Phi_n(x) = \sup_{n \leq k} f_k(x)$$

for each $n \in \mathbb{N}$ and every $x \in X$. Since each f_n is measurable, the foregoing argument ensures that the functions ϕ_n and Φ_n also are measurable, and so are the functions $\underline{f}$ and $\overline{f}$ (by the same argument). Indeed,

$$\underline{f}(x) = \sup_n \inf_{n \leq k} f_k(x) = \sup_n \phi_n(x) \quad \text{and} \quad \overline{f}(x) = \inf_n \sup_{n \leq k} f_k(x) = \inf_n \Phi_n(x)$$

for each $x \in X$. Finally, recall that f exists if and only if $\underline{f} = \overline{f}$. $\square$

We are now ready to prove the extension of Propositions 1.5 and 1.6 to extended real-valued functions. But first we need to pose some conventions. First we declare that $0 \cdot (\pm\infty) = 0$ so that, if $\gamma = 0$, then $\gamma f(x) = 0$ for all $x \in X$ (i.e., $\gamma f = 0$ whenever $\gamma = 0$) for every $\overline{\mathbb{R}}$-valued function f on X. Moreover, if f and g are $\overline{\mathbb{R}}$-valued $\mathcal{X}$-measurable functions on X, then

$$F_{+\infty} \cap G_{-\infty} = \big\{ x \in X \colon f(x) = +\infty \ \text{and} \ g(x) = -\infty \big\}$$

and

$$G_{+\infty} \cap F_{-\infty} = \big\{ x \in X \colon g(x) = +\infty \ \text{and} \ f(x) = -\infty \big\}$$

are $\mathcal{X}$-measurable sets (Problem 1.4), but the function $f + g$ is not pointwise defined on these sets (i.e., $f + g$ is not defined by the formula $(f + g)(x) = f(x) + g(x)$ on $F_{+\infty} \cap G_{-\infty}$ or $G_{+\infty} \cap F_{-\infty}$). Thus we declare

$$(f + g)(x) = 0 \quad \text{for all} \quad x \in (F_{+\infty} \cap G_{-\infty}) \cup (G_{+\infty} \cap F_{-\infty}).$$

Finally, recall the definition of the functions $f \wedge g$ and $f \vee g$: for each $x \in X$, $(f \wedge g)(x) = \min\{f(x), g(x)\}$ and $(f \vee g)(x) = \max\{f(x), g(x)\}$.

Proposition 1.9. *If f and g are $\overline{\mathbb{R}}$-valued measurable, then so are*

$$\gamma f, \quad f + g, \quad f^+, \quad f^-, \quad |f|, \quad f \wedge g, \quad f \vee g \quad \text{and} \quad fg.$$

Proof. Let f and g be $\mathcal{X}$-measurable $\overline{\mathbb{R}}$-valued functions and let γ be any *real* number. The previous conventions ensure that the functions γf and $f + g$ are well defined, and they are $\mathcal{X}$-measurable by using the same argument in the proof of Proposition 1.5. The arguments in the first part of the proof of Proposition 1.6 still hold for $\overline{\mathbb{R}}$-valued functions, so f^+, f^-, and $|f|$ are, again, well-defined $\mathcal{X}$-measurable functions. Similarly (see Problem 1.3), the functions $f \wedge g$ and $f \vee g$ also are well defined and $\mathcal{X}$-measurable.

Note that fg is a well-defined $\overline{\mathbb{R}}$-valued function. To verify that it is $\mathcal{X}$-measurable, proceed as follows. For each pair of integers $m, n \in \mathbb{N}$, consider the truncate functions $f_n\colon X \to \mathbb{R}$ and $g_m\colon X \to \mathbb{R}$ given by

$$f_n(x) = \begin{cases} f(x), & |f(x)| \leq n, \\ n, & f(x) > n, \\ -n, & f(x) < -n, \end{cases} \quad \text{and} \quad g_m(x) = \begin{cases} g(x), & |g(x)| \leq m, \\ m, & g(x) > m, \\ -m, & g(x) < -m, \end{cases}$$

for every $x \in X$. Recall that $\{f_n\}$ and $\{g_m\}$ are sequences of $\mathbb{R}$-valued $\mathcal{X}$-measurable functions (cf. Problem 1.5), and so the functions $f_n\,g_m$ are $\mathbb{R}$-valued $\mathcal{X}$-measurable for each pair of integers $m, n \in \mathbb{N}$ according to Proposition 1.5. It is readily verified that the sequences $\{f_n\}$ and $\{g_m\}$ converge pointwise to f and g, respectively. Thus, since $\{f_n\,g_m\}$ is a sequence of $\mathcal{X}$-measurable functions for each $m \in \mathbb{N}$, and since

$$f g_m = \lim_n f_n\,g_m$$

for each $m \in \mathbb{N}$, it follows that $f g_m$ is $\mathcal{X}$-measurable for each $m \in \mathbb{N}$ by Proposition 1.8. Since each function $f g_m$ is $\mathcal{X}$-measurable, and since

$$fg = \lim_m f g_m,$$

it follows, again by Proposition 1.8, that fg is $\mathcal{X}$-measurable. $\square$

Remark: We can use the argument in the proof of Proposition 1.5 to see that the function f^2 is well defined and $\mathcal{X}$-measurable, but the above conventions are not enough to ensure the identity $fg = \frac{1}{4}((f+g)^2 - (f-g)^2)$ — Why?

The collection of all extended real-valued functions on X, measurable with respect to a σ-algebra $\mathcal{X}$ of subsets of X, will be denoted by $\mathcal{M}(X, \mathcal{X})$ or simply by $\mathcal{M}$ when the measurable space $(X, \mathcal{X})$ is clear in the context:

$$\mathcal{M} = \mathcal{M}(X, \mathcal{X}) = \{f\colon X \to \overline{\mathbb{R}}\colon f \text{ is } \mathcal{X}\text{-measurable}\}.$$

1.4   Problems

Problem 1.1. Let $(X, \mathcal{X})$ be a measurable space and $f\colon X \to \mathbb{R}$ a real-valued function on X. Show that the assertions below are pairwise equivalent.

(a) $f^{-1}((\alpha, \infty)) = \{x \in X\colon f(x) > \alpha\} \in \mathcal{X}$ for every $\alpha \in \mathbb{R}$.

(b) $f^{-1}([\alpha, \infty)) = \{x \in X\colon f(x) \geq \alpha\} \in \mathcal{X}$ for every $\alpha \in \mathbb{R}$.

(c) $f^{-1}((-\infty, \alpha)) = \{x \in X\colon f(x) < \alpha\} \in \mathcal{X}$ for every $\alpha \in \mathbb{R}$.

(d) $f^{-1}\big((-\infty,\alpha]\big) = \{x \in X\colon f(x) \le \alpha\} \in \mathcal{X}$ for every $\alpha \in \mathbb{R}$.

Now show that these four assertions can be equivalently stated if we replace "for every $\alpha \in \mathbb{R}$" with "for every $\alpha \in \mathbb{Q}$".

Hint: If $\alpha \in \mathbb{R}$, then $\alpha = \lim_n \alpha_n$ with $\alpha > \alpha_n > \alpha_{n+1} \in \mathbb{Q}$ for every $n \ge 1$ so that $f^{-1}\big((\alpha,\infty)\big) = \bigcup_n \{x \in X\colon f(x) > \alpha_n\}$.

Thus all the preceding eight assertions are equivalent forms of defining a measurable function, and they still hold if $f\colon X \to \overline{\mathbb{R}}$ take values in $\overline{\mathbb{R}}$.

Problem 1.2. Take the real line $\mathbb{R}$ and the Borel algebra $\mathfrak{R}$. All functions in this problem are real-valued on $\mathbb{R}$, and by a measurable one we mean an $\mathfrak{R}$-measurable (or *Borel measurable*) function. Prove the following assertions.

(a) Every continuous function is measurable.

(b) Every monotone function is measurable.

Now consider a function (the so-called *Dirichlet function*) that is far from being continuous or monotone.

(c) The characteristic function $\mathcal{X}_\mathbb{Q}$ of the rationals is measurable.

Problem 1.3. Suppose $f\colon X \to \mathbb{R}$ and $g\colon X \to \mathbb{R}$ are measurable functions (with respect to a σ-algebra $\mathcal{X}$ of subsets of X). Consider the functions $e\colon X \to \mathbb{R}$ and $h\colon X \to \mathbb{R}$ defined, for each $x \in X$, by

$$e(x) = \min\big\{f(x),g(x)\big\} \quad \text{and} \quad h(x) = \max\big\{f(x),g(x)\big\}$$

(notations: $e = f \wedge g = \inf\{f,g\}$ and $h = f \vee g = \sup\{f,g\}$). Show that

$$e = \tfrac{1}{2}\big(f + g - |f - g|\big) \quad \text{and} \quad h = \tfrac{1}{2}\big(f + g + |f - g|\big).$$

Now conclude: e and h are measurable functions ($\mathcal{X}$-measurable, that is).

Problem 1.4. Let $f\colon X \to \overline{\mathbb{R}}$ be an extended real-valued function. Show that, if f is measurable with respect to a σ-algebra $\mathcal{X}$ of subsets of X, then

$$F_{+\infty} = \{x \in X\colon f(x) = +\infty\} \quad \text{and} \quad F_{-\infty} = \{x \in X\colon f(x) = -\infty\}$$

are $\mathcal{X}$-measurable sets. *Hint:* For n ranging over $\mathbb{N}$,

$$F_{+\infty} = \bigcap_n \{x \in X\colon f(x) > n\}$$

and

$$F_{-\infty} = \bigcap_n \{x \in X\colon f(x) \le -n\} = X \backslash \bigcup_n \{x \in X\colon f(x) > -n\};$$

countable intersection and complement of countable union.

Problem 1.5. Let $(X, \mathcal{X})$ be a measurable space and let $f : X \to \mathbb{R}$ be a measurable function. For each *positive real* number β define the *β-truncation* of f as the function $f_\beta : X \to \mathbb{R}$ given, for each $x \in X$, by

$$
f_\beta(x) = \begin{cases} f(x), & |f(x)| \leq \beta, \\ \beta, & f(x) > \beta, \\ -\beta, & f(x) < -\beta. \end{cases}
$$

Verify that f_β is a measurable function.

Problem 1.6. Let $f : X \to \mathbb{R}$ be a nonnegative $\mathcal{X}$-measurable function, where $\mathcal{X}$ is a σ-algebra of subsets of X. Show that there exists a sequence $\{\varphi_n\}$ of $\mathcal{X}$-measurable functions $\varphi_n : X \to \mathbb{R}$ with the following properties.

(i) Each φ_n is nonnegative (i.e., $0 \leq \varphi_n(x)$ for every $x \in X$ for each n).

(ii) $\{\varphi_n\}$ is increasing (i.e., $\varphi_n(x) \leq \varphi_{n+1}(x)$ for every $x \in X$ for each n).

(iii) $\{\varphi_n\}$ converges pointwise to f (i.e., $f(x) = \lim_n \varphi_n(x)$ for every $x \in X$).

(iv) Each φ_n has a finite range (i.e., $\{\alpha \in \mathbb{R} : \alpha = \varphi_n(x) \text{ for some } x \in X\}$ is a finite set for each n).

Also show that if f is *bounded* (i.e., if $\sup_{x \in X} |f(x)| < \infty$), then $\{\varphi_n\}$ *converges uniformly* to f (i.e., $\sup_{x \in X} |\varphi_n(x) - f(x)| \to 0$ as $n \to \infty$).

Hint: Take an arbitrary $n \in \mathbb{N}$ and, for each integer $0 \leq k \leq n2^n$, put

$$
E_{n,k} = \begin{cases} \{x \in X : k2^{-n} \leq f(x) < (k+1)2^{-n}\}, & k < n2^n, \\ \{x \in X : f(x) \geq n\}, & k = n2^n, \end{cases}
$$

so that $\{E_{n,k}\}_{0 \leq k \leq n2^n}$ is a partition of X made up of $\mathcal{X}$-measurable sets. Now, for each $n \in \mathbb{N}$, put $\varphi_n(x) = 2^{-n} \sum_{k=0}^{n2^n} k \chi_{E_{n,k}}(x)$ for every $x \in X$.

Problem 1.7. Let $f : X \to \mathbb{C}$ be any complex-valued function on X. Recall that f can be uniquely decomposed as $f = f_1 + i f_2$, where f_1 and f_2 are real-valued functions on X (called the real and imaginary parts of f and defined by $f_1(x) = \operatorname{Re} f(x)$ and $f_2(x) = \operatorname{Im} f(x)$ for every $x \in X$). We say that a complex-valued function f is *measurable* (with respect to the σ-algebra $\mathcal{X}$), or $\mathcal{X}$-*measurable*, if its real and imaginary parts f_1 and f_2 are both (real-valued) $\mathcal{X}$-measurable functions.

(a) Show that $f : X \to \mathbb{C}$ is $\mathcal{X}$-measurable if and only if the inverse image of every open rectangle of the complex plane $\mathbb{C}$ is an $\mathcal{X}$-measurable set. That is, the set $\{x \in X : \alpha < f_1(x) < \beta \text{ and } \gamma < f_2(x) < \delta\}$ lies in $\mathcal{X}$ for all real numbers α, β, γ, and δ.

(b) Generalize: show that $f: X \to \mathbb{C}$ is $\mathcal{X}$-measurable if and only if the inverse image of every open set of $\mathbb{C}$ is an $\mathcal{X}$-measurable set.

(c) Show that sums and products of complex-valued measurable functions are again measurable as well as the limit of every (pointwise) convergent sequence of complex-valued measurable functions.

Problem 1.8. Let $(X, \mathcal{X})$ and $(Y, \mathcal{Y})$ be measurable spaces and consider a function F of X into Y. We say that F is a *measurable transformation* (with respect to the σ-algebras $\mathcal{X}$ and $\mathcal{Y}$) if the inverse image of every $\mathcal{Y}$-measurable set is an $\mathcal{X}$-measurable set. That is, $F: X \to Y$ is measurable with respect to the σ-algebras $\mathcal{X}$ and $\mathcal{Y}$ of subsets of X and Y if

$$F^{-1}(E) \in \mathcal{X} \quad \text{for every} \quad E \in \mathcal{Y}.$$

(Note the similarity with the definition of continuous function between topological spaces.) Now put $Y = \mathbb{R}$, $\mathcal{Y} = \Re$ (the Borel algebra) and consider a real-valued function $f: X \to \mathbb{R}$. Show that f is $\mathcal{X}$-measurable in the sense of Definition 1.2 if and only if it is measurable in the above sense (i.e., $f^{-1}(E) \in \mathcal{X}$ for every Borel set E).

Problem 1.9. Take any function F of a set X into a set Y. Let $\mathcal{Y}$ be a σ-algebra of subsets of Y. Show that the collection of the inverse images under F of each $\mathcal{Y}$-measurable set, $\mathcal{X} = \{F^{-1}(E): E \in \mathcal{Y}\}$, forms a σ-algebra of subsets of X. Moreover, given a function $F: X \to Y$ and a σ-algebra $\mathcal{Y}$ of subsets of Y, verify that this $\mathcal{X}$ is the smallest σ-algebra of subsets of X that makes F measurable, which is called the σ-algebra of subsets of X *inversely induced* by F. (Note the analogy between this concept and that of the topology inversely induced on X by F, which is the weakest topology on X that makes F continuous.)

Hint: Take an arbitrary function $F: X \to Y$ and recall that the *inverse image* of any subset B of Y under F is the subset of X given by

$$F^{-1}(B) = \{x \in X: F(x) \in B\}.$$

Verify that $\varnothing = F^{-1}(\varnothing)$, $X = F^{-1}(Y)$, $X \backslash F^{-1}(B) = F^{-1}(Y \backslash B)$ for every $B \subseteq Y$, and that $\bigcup_\gamma F^{-1}(B_\gamma) = F^{-1}(\bigcup_\gamma B_\gamma)$ for every nonempty collection $\{B_\gamma\}$ of subsets of Y.

Problem 1.10. Let $(X, \mathcal{X})$, $(Y, \mathcal{Y})$, and $(Z, \mathcal{Z})$ be three measurable spaces. Suppose $F: X \to Y$ and $G: Y \to Z$ are measurable functions (with respect to the σ-algebras $\mathcal{X}$ and $\mathcal{Y}$, and $\mathcal{Y}$ and $\mathcal{Z}$, respectively). Show that the composition $G \circ F: X \to Z$ is measurable (with respect to the σ-algebras $\mathcal{X}$ and $\mathcal{Z}$).

Problem 1.11. Let $(X, \mathcal{X})$ be a measurable spaces. Suppose $f: X \to \mathbb{R}$ is an $\mathcal{X}$-measurable function and $g: \mathbb{R} \to \mathbb{R}$ is a continuous function. Show that the composition $g \circ f: X \to \mathbb{R}$ is measurable. (*Hint:* Problems 1.2 and 1.10.)

Problem 1.12. A collection $\mathcal{T}$ of subsets of a set X is a *topology* on X if it satisfies the following axioms.

(i) The whole set X and the empty set $\varnothing$ belong to $\mathcal{T}$.

(ii) The intersection of a finite collection of sets in $\mathcal{T}$ belongs to $\mathcal{T}$.

(iii) The union of an arbitrary collection of sets in $\mathcal{T}$ belongs to $\mathcal{T}$.

A set X equipped with a topology $\mathcal{T}$ is referred to as a *topological space*, denoted by $(X, \mathcal{T})$ or simply by X (if $\mathcal{T}$ is clear or immaterial). The sets in $\mathcal{T}$ are called the *open* sets of X with respect to $\mathcal{T}$. Let $\mathcal{X}_\mathcal{T}$ be the σ-algebra generated by a topology $\mathcal{T}$ on a set X. Show that if a real-valued function $f: X \to \mathbb{R}$ is continuous (with respect to the topology $\mathcal{T}$, which means that the inverse image under f of every open set of the real line is an open set of X), then f is measurable (with respect to the σ-algebra $\mathcal{X}_\mathcal{T}$).

Problem 1.13. Generalize the result of Problem 1.12. Let $\mathcal{X}_\mathcal{T}$ be the σ-algebra generated by a topology $\mathcal{T}_X$ on a set X and let $\mathcal{Y}_\mathcal{T}$ be the σ-algebra generated by a topology $\mathcal{T}_Y$ on a set Y. If $F: X \to Y$ is a *continuous transformation* (i.e., if $F^{-1}(U) \in \mathcal{T}_X$ for every $U \in \mathcal{T}_Y$), then F is measurable. (Note that "continuity" is with respect to the topologies $\mathcal{T}_X$ and $\mathcal{T}_Y$, while "measurability" is with respect to the σ-algebras $\mathcal{X}_\mathcal{T}$ and $\mathcal{Y}_\mathcal{T}$.)

Problem 1.14. Let X be a topological space. A subcollection $\mathcal{B}$ of the topology $\mathcal{T}$ is a *base* (or a *topological base*) for X if it covers every open subset of X (i.e., every $U \in \mathcal{T}$ is the union of some subcollection of $\mathcal{B}$). If X is a metric space equipped with the metric topology $\mathcal{T}$, then it has a base of open balls (every open set is the union of open balls). A topological space is *separable* if it has a countable base. A metric space is separable if and only if it has a countable base of open balls. Show that the σ-algebra $\mathcal{X}_\mathcal{T}$ generated by the metric topology $\mathcal{T}$ of a separable metric space coincides with the σ-algebra $\mathcal{X}_\mathcal{B}$ generated by any countable base $\mathcal{B}$ of open balls.

Remark: The real line $\mathbb{R}$ (equipped with its usual topology) is a separable metric space. Thus the Borel algebra $\mathfrak{R}$ (the σ-algebra generated by the open intervals) coincides with the σ-algebra generated by any countable base of open intervals, which in turn coincide with the σ-algebra generated by the topology of $\mathbb{R}$ (the σ-algebra generated by the open sets of $\mathbb{R}$). For this reason $\mathcal{X}_\mathcal{T}$ is referred to as the *Borel σ-algebra* of subsets of X.

Problem 1.15. A *monotone class* is a nonempty class (i.e., a nonempty collection) $\mathcal{K}$ of subsets of a set X that contains the union of every increasing sequence in $\mathcal{K}$ and the intersection of every decreasing sequence in $\mathcal{K}$. That is, if $\{E_n\}$ is an increasing sequence ($E_n \subseteq E_{n+1}$) of sets in $\mathcal{K}$ and $\{F_n\}$ is a decreasing sequence ($F_{n+1} \subseteq F_n$) of sets in $\mathcal{K}$, then $\bigcup_n E_n$ and $\bigcap_n F_n$ are sets in $\mathcal{K}$. The remaining problems are all about monotone classes, leading to Problem 1.18, which in turn will be required later in the sequel. To begin with, prove the following assertions.

   (a) Every σ- algebra is a monotone class.

   (b) A monotone class is not necessarily a σ-algebra.

A *monotone algebra* is a nonempty collection of subsets of a set X that is both a monotone class and an algebra.

   (c) Every monotone algebra is a σ-algebra.

   Hint: Let $\mathcal{K}$ be a monotone algebra and let $\{E_n\}$ be a sequence of sets in $\mathcal{K}$. Since $\mathcal{K}$ is an algebra, $\{\bigcup_{i=1}^{n} E_i\}$ is an increasing sequence of sets in $\mathcal{K}$. Since $\mathcal{K}$ is a monotone class, $\{\bigcup_{i=1}^{\infty} E_i\}$ lies in $\mathcal{K}$.

Problem 1.16. If $\mathcal{C}$ is a nonempty collection of subsets of X, then there is a smallest (inclusion ordering) monotone class $\mathcal{K}_\mathcal{C}$ of subsets of X that includes $\mathcal{C}$. That is, $\mathcal{K}_\mathcal{C}$ is included in any monotone class that includes $\mathcal{C}$.

Hint: The intersection of any collection of monotone classes of subsets of X is again a monotone class of subsets of X.

This smallest monotone class $\mathcal{K}_\mathcal{C}$ is the *monotone class generated by $\mathcal{C}$.*

Problem 1.17. Show that the monotone class $\mathcal{K}_\mathcal{C}$ generated by a collection $\mathcal{C}$ of subsets of X is included in the σ-algebra $\mathcal{X}_\mathcal{C}$ generated by $\mathcal{C}$:

$$\mathcal{C} \subseteq \mathcal{K}_\mathcal{C} \subseteq \mathcal{X}_\mathcal{C}.$$

Moreover, give an example where the foregoing inclusions are all proper.

Problem 1.18. However, if the collection is an algebra $\mathcal{A}$, then

$$\mathcal{K}_\mathcal{A} = \mathcal{X}_\mathcal{A}.$$

That is, prove that the monotone class $\mathcal{K}_\mathcal{A}$ generated by an algebra $\mathcal{A}$ of subsets of a set X coincides with the σ-algebra $\mathcal{X}_\mathcal{A}$ generated by $\mathcal{A}$. This is the *Monotone Class Lemma*, which plays a crucial role in Chapter 9.

Hint: Let $\mathcal{A}$ be an algebra of subsets of a set X and let $\mathcal{K}_\mathcal{A}$ be the monotone class generated by $\mathcal{A}$. For each $E \in \mathcal{K}_A$, put

$$\mathcal{E}_A = \{F \in \mathcal{K}_A : E\backslash F \in \mathcal{K}_A,\ E \cap F \in \mathcal{K}_A \text{ and } F\backslash E \in \mathcal{K}_A\} \subseteq \mathcal{K}_A.$$

First show that

(i) $\mathcal{E}_A$ is a monotone class,

(ii) $F \in \mathcal{E}_A$ if and only if $E \in \mathcal{F}_A$,

where the definition of $\mathcal{F}_A$ is analogous to that of $\mathcal{E}_A$ (exchanging E for F). Since $\mathcal{A}$ is an algebra and $\mathcal{A} \subseteq \mathcal{K}_A$ (cf. Problem 1.17), verify that

(iii) $\varnothing$, E, $X\backslash E$, and X all lie in $\mathcal{E}_A$,

(iv) $E \in \mathcal{A}$ implies $\mathcal{A} \subseteq \mathcal{E}_A$.

(Recall: $E \cap F = X\backslash((X\backslash E) \cup (X\backslash F))$ and $E\backslash F = E \cap (X\backslash F)$ so that $E\backslash F$, $E \cap F$, and $F\backslash E$ lie in $\mathcal{A}$ whenever E and F lie in $\mathcal{A}$.) Now conclude from (i) and (iv) that (cf. Problem 1.16)

$$\mathcal{E}_A = \mathcal{K}_A \quad \text{for every} \quad E \in \mathcal{A}.$$

Hence, if $E \in \mathcal{A}$ and $F \in \mathcal{K}_A$, then $F \in \mathcal{E}_A$, and so $E \in \mathcal{F}_A$ by (ii). Thus $\mathcal{A} \subseteq \mathcal{F}_A$ for every $F \in \mathcal{K}_A$. Then conclude from (i) that (cf. Problem 1.16)

$$\mathcal{F}_A = \mathcal{K}_A \quad \text{for every} \quad F \in \mathcal{K}_A.$$

Therefore, if $E, F \in \mathcal{K}_A$, then $E \cap F$, $E\backslash F$, and $F\backslash E$ lie in $\mathcal{K}_A$. Moreover, it follows by (iii) that $\varnothing$ and X also lie in $\mathcal{K}_A$. Thus conclude that intersection and complement of sets in $\mathcal{K}_A$ remain in $\mathcal{K}_A$, and so a finite union of sets in $\mathcal{K}_A$ remain in $\mathcal{K}_A$ (recall that $F \cup E = X\backslash((X\backslash E) \cap (X\backslash F))$). Hence, according to Definition 1.1,

$$\mathcal{K}_A \text{ is an algebra.}$$

Apply Problem 1.15(c) to verify that $\mathcal{K}_A$ is a σ-algebra, and hence infer that $\mathcal{X}_A \subseteq \mathcal{K}_A$. Finally, use Problem 1.17 to conclude that $\mathcal{K}_A = \mathcal{X}_A$.

Problem 1.19. Show that if a monotone class includes an algebra $\mathcal{A}$, then it also includes the σ-algebra $\mathcal{X}_A$ generated by $\mathcal{A}$.

Suggested Reading

Bartle [3], Berberian [6], Halmos [13], Royden [22], Rudin [23]. For an introduction to set theory, see [12], [25], [26] and the first chapter of [8], [9], [13], [14], [16], [17], [18], [22]. For general topology, see [8], [9], [14], [16, Chapters 2 & 3], [17, Chapter 3], [18, Chapters 7 & 8], [22, Part Two].

2

Measure

2.1 Measure Space

A *set function* is a function whose domain is a collection of sets. A measure is a special kind of nonnegative extended real-valued set function. Let us assume that the domain of a measure is a subcollection of the power set of a given fixed set. It is convienient to force the empty set and the whole set itself into the domain and attribute the minimum (zero) for the value of the function at the empty set. It is also convienient to require *additivity*, in the following sense. First we must assume that any finite union of sets in the domain is again a set in the domain. This suggests that the domain might be an algebra. Then we require that the value of the function at any finite union of disjoint sets in the domain equals the sum of the values of the function at each set. Indeed, this might be a good start for the definition of a concept of measure, but it lacks an important feature that will be needed to build up a useful theory, namely, *countable additivity*. That is, we shall need that additivity also holds for countably infinite unions of disjoint sets. This forces the domain to be more than an algebra; a σ-algebra is enough.

Definition 2.1. Let $\mathcal{X}$ be a σ-algebra of subsets of a set X. A function

$$\mu\colon \mathcal{X} \to \overline{\mathbb{R}}$$

(i.e., an extended real-valued set function μ on $\mathcal{X}$) is a *measure* if it satisfies the following conditions (which are called the *measure axioms*).

   (a) $\mu(\varnothing) = 0$.

(b) $\mu(E) \geq 0$ for every $E \in \mathcal{X}$.

(c) $\mu\left(\bigcup_n E_n\right) = \sum_n \mu(E_n)$

for every countable family $\{E_n\}$ of pairwise disjoint $(E_m \cap E_n = \varnothing$ whenever $m \neq n)$ sets in $\mathcal{X}$.

By a *measure space* we mean a triple $(X, \mathcal{X}, \mu)$, where X is an arbitrary set, $\mathcal{X}$ is a σ-algebra of subsets of X, and μ is a measure on $\mathcal{X}$.

Recall that μ takes values in $\overline{\mathbb{R}}$ so that there may be $\mathcal{X}$-measurable sets E for which $\mu(E) = +\infty$. If the countable family $\{E_n\}$ is infinite and all the sets E_n have a finite measure (i.e., if $\mu(E_n) \in \mathbb{R}$ for every n), then we have an infinite sum of nonnegative numbers in (c), which means either a convergent (unconditionally convergent, actually) or divergent series of nonnegative numbers. If the series diverges, then $\mu\left(\bigcup_n E_n\right) = +\infty$.

A *finite measure* is a real-valued measure $\mu \colon \mathcal{X} \to \mathbb{R}$. This is expressed by writing $\mu(E) < \infty$ for all $E \in \mathcal{X}$ or, equivalently, $\mu(X) < \infty$ (cf. upcoming Proposition 2.2(a)). In particular, if $\mu(X) = 1$, then μ is called a *probability measure* and $(X, \mathcal{X}, \mu)$ a *probability space*. If there exists a countable covering of X consisting $\mathcal{X}$-measurable sets of finite measure, then μ is a σ-*finite measure*. In other words, a measure $\mu \colon \mathcal{X} \to \overline{\mathbb{R}}$ is σ-*finite* if there exists an $\mathcal{X}$-valued sequence $\{E_n\}$ such that $\mu(E_n) < \infty$ for every n and $X = \bigcup_n E_n$.

Example 2A. Let $\mathcal{X}$ be a σ-algebra of subsets of a nonempty set X. For each $x \in X$ consider the function $\delta_x \colon \mathcal{X} \to \mathbb{R}$ defined, for each $E \in \mathcal{X}$, by

$$\delta_x(E) = \begin{cases} 1, & x \in E, \\ 0, & x \in X \backslash E. \end{cases}$$

This is a probability measure, referred to as the *unit point measure* concentrated at x (or the *Dirac measure* at x). An *atom* of a measure μ on $\mathcal{X}$ is a measurable set A such that $\mu(A) > 0$ and, for every measurable subset E of A, either $\mu(E) = 0$ or $\mu(E) = \mu(A)$. Note that any measurable set containing x is an atom of δ_x. A *singleton* is a set containing just one element. If singletons are measurable with respect to $\mathcal{X}$, then $\{x\}$ is an atom of δ_x, and $\delta_x(\{y\})$ is 1 if $y = x$ and 0 if $y \neq x$ for every singleton $\{y\}$ in $\mathcal{X}$.

Example 2B. Take the σ-algebra $\wp(\mathbb{N})$ of all subsets of the positive integers $\mathbb{N}$. Let $\#$ stand for cardinality, which, for any *finite* set, means the number of elements in it. Consider the function $\mu \colon \wp(\mathbb{N}) \to \overline{\mathbb{R}}$ defined by

$$\mu(E) = \begin{cases} \#(E), & E \text{ is finite}, \\ +\infty, & E \text{ is infinite}, \end{cases}$$

for every $E \in \wp(\mathbb{N})$. It is easy to check that μ satisfies all the axioms of Definition 2.1. Now put $E_n = \{n\} \in \wp(\mathbb{N})$ for each $n \in \mathbb{N}$ so that $\mathbb{N} = \bigcup_n E_n$ and $\mu(E_n) = 1$. Thus μ is a σ-finite measure but not a finite measure (there are infinite sets in $\wp(\mathbb{N})$). This is called the *counting measure* on $\mathbb{N}$.

Example 2C. Let $\Re$ be the Borel algebra, which is the σ-algebra of subsets of the real line $\mathbb{R}$ generated by the collection of all open intervals. We shall verify later the existence and uniqueness of a measure $\lambda: \Re \to \overline{\mathbb{R}}$ such that $\lambda((\alpha, \beta)) = \beta - \alpha$ for every open interval (α, β) of $\mathbb{R}$ (Chapter 8). That is, this measure is the unique measure on $\Re$ that has the property of assigning to each open interval its own length. It is called the *Lebesgue measure* on $\Re$, which is not finite (for instance, $\lambda(\mathbb{R}) = +\infty$) but is σ-finite. Indeed, take an arbitrary $\varepsilon > 0$ and, for each integer $k \in \mathbb{Z}$, let $E_k = (q_k - \varepsilon, q_k + \varepsilon)$ be the open interval of radius ε centered at q_k, where $\{q_k\}_{k \in \mathbb{Z}}$ is an enumeration of the rational numbers $\mathbb{Q}$, so that $\mathbb{R} = \bigcup_k E_k$ and $\mu(E_k) = 2\varepsilon$ for all $k \in \mathbb{Z}$.

Example 2D. Consider the previous example. Observe that scaling the Lebesgue measure λ trivially yields another measure on $\Re$. That is, for any $\gamma > 0$, the function $\lambda_\gamma: \Re \to \overline{\mathbb{R}}$ defined by $\lambda_\gamma = \gamma\lambda$ (so that $\lambda_\gamma((\alpha, \beta)) = \gamma\lambda((\alpha, \beta)) = \gamma(\beta - \alpha) = \gamma\beta - \gamma\alpha$ for every open interval (α, β) of $\mathbb{R}$) is again a measure on $\Re$. This is a homogeneous scaling. However, inhomogeneous scaling may also yield new measures as follows. Let $f: \mathbb{R} \to \mathbb{R}$ be any continuous increasing function (i.e., f is continuous and $f(x) \le f(y)$ whenever $x \le y$). The arguments used to build up the Lebesgue measure λ (Chapter 8) can be easily modified to show that there exists a unique measure $\lambda_f: \Re \to \overline{\mathbb{R}}$ such that $\lambda_f((\alpha, \beta)) = f(\beta) - f(\alpha)$ for every open interval (α, β) of the real line $\mathbb{R}$. This is the *Borel–Stieltjes measure* generated by f. The same argument of Example 2C ensures that λ_f is σ-finite.

Recall that a sequence $\{A_n\}$ of subsets of X is *increasing* if $A_n \subseteq A_{n+1}$ for every n and *decreasing* if $A_{n+1} \subseteq A_n$ for every n. The next proposition presents some useful basic properties of a measure function.

Proposition 2.2. *Consider a σ-algebra $\mathcal{X}$ of subsets of a set X. Let A and B be $\mathcal{X}$-measurable sets and let $\{E_n\}$ be a sequence of sets in $\mathcal{X}$. The following properties hold true for any measure $\mu: \mathcal{X} \to \overline{\mathbb{R}}$.*

(a) $\mu(A) \le \mu(B)$ *if* $A \subseteq B$.

(b) $\mu(B \backslash A) = \mu(B) - \mu(A)$ *if* $A \subseteq B$ *and* $\mu(A) \ne +\infty$.

(c) $\mu\left(\bigcup_n E_n\right) = \lim_n \mu(E_n)$ *if* $\{E_n\}$ *is increasing*.

(d) $\mu\left(\bigcap_n E_n\right) = \lim_n \mu(E_n)$ *if* $\{E_n\}$ *is decreasing and* $\mu(E_1) \ne +\infty$.

Proof. Recall that if $A \subseteq B \subseteq X$, then $B = A \cup (B \backslash A)$ and $A \cap (B \backslash A) = \varnothing$. Moreover, if both A and B are $\mathcal{X}$-measurable, then so is $B \backslash A = B \cap (X \backslash A)$ (Definition 1.1(b,c)). Therefore, according to Definition 2.1(b,c)

$$\mu(A) \leq \mu(A) + \mu(B \backslash A) = \mu(A \cup (B \backslash A)) = \mu(B).$$

Moreover, if in addition $\mu(A) \neq +\infty$, then $\mu(B) - \mu(A)$ lies in $\mathbb{R}$ so that

$$\mu(B) - \mu(A) = \mu(B \backslash A).$$

This proves (a) and (b). Now suppose $\{E_n\}$ is an increasing sequence of $\mathcal{X}$-measurable sets. If one of them has an infinite measure, then assertion (c) is verified by (a). Thus suppose $\mu(E_n) \in \mathbb{R}$ for every $n \in \mathbb{N}$, and consider a sequence $\{E_n'\}$ of $\mathcal{X}$-measurable sets recursively defined as follows: $E_1' = E_1$ and $E_{n+1}' = E_{n+1} \backslash E_n$ for each $n \in \mathbb{N}$, which is a sequence of pairwise disjoint sets, and hence (cf. Definition 2.1(c) and recall that $\bigcup_n E_n = \bigcup_n E_n'$)

$$\mu\Big(\bigcup_n E_n\Big) = \mu\Big(\bigcup_n E_n'\Big) = \sum_n \mu(E_n') = \mu(E_1') + \lim_m \sum_{n=1}^m \mu(E_{n+1}').$$

Now, for an arbitrary integer $m > 1$, it follows from (b) that

$$\sum_{n=1}^m \mu(E_{n+1}') = \sum_{n=1}^m \mu(E_{n+1} \backslash E_n)$$
$$= \sum_{n=1}^m \big(\mu(E_{n+1}) - \mu(E_n)\big) = \mu(E_{m+1}) - \mu(E_1),$$

and so $\mu\big(\bigcup_n E_n\big) = \lim_m \mu(E_{m+1})$, which proves (c). Finally, suppose $\{E_n\}$ is a decreasing sequence of $\mathcal{X}$-measurable sets and put $E_n'' = E_1 \backslash E_n$ for each n so that $\{E_n''\}$ is an increasing sequence, and so we can apply (c). But first observe that, since $E_n \subseteq E_1$ for all n and $\mu(E_1) \neq +\infty$, we get from (a) that $\mu(E_n) \in \mathbb{R}$ for all n (and so $\mu\big(\bigcap_n E_n\big) \in \mathbb{R}$). Thus, according to (b),

$$\mu(E_1 \backslash E_n) = \mu(E_1) - \mu(E_n) \quad \text{and} \quad \mu\Big(E_1 \backslash \bigcap_n E_n\Big) = \mu(E_1) - \mu\Big(\bigcap_n E_n\Big)$$

for every n. Therefore, recalling the De Morgan laws and applying (c),

$$\mu(E_1) - \mu\Big(\bigcap_n E_n\Big) = \mu\Big(E_1 \backslash \bigcap_n E_n\Big) = \mu\Big(\bigcup_n (E_1 \backslash E_n)\Big) = \mu\Big(\bigcup_n E_n''\Big)$$
$$= \lim_n \mu(E_n'') = \lim_n \mu(E_1 \backslash E_n) = \mu(E_1) - \lim_n \mu(E_n),$$

which completes the proof of (d) once $\mu(E_1) \neq +\infty$. $\square$

Let $(X, \mathcal{X}, \mu)$ be a measure space. If a statement $P(x)$ holds for every $x \in X \backslash N$ for some $N \in \mathcal{X}$ such that $\mu(N) = 0$, then we say that $P(x)$ holds

μ-*almost everywhere* (or *almost everywhere with respect to* μ, or simply *almost everywhere* if the measure μ is clear in the context, or still *almost sure* if $(X, \mathcal{X}, \mu)$ is a probability space). That is, if $P(x)$ is true up to a set of measure zero, which means that there exists an $\mathcal{X}$-measurable set N with $\mu(N) = 0$ such that $P(x)$ holds true on the complement on N.

Example 2E. Consider a couple of functions $f: X \to Y$ and $g: X \to Y$ from a set X to a set Y. Recall the usual definition of equality between functions: $f = g$ if $f(x) = g(x)$ for every $x \in X$. That is, equality is pointwise defined everywhere in X. Now suppose $\mu: \mathcal{X} \to \overline{\mathbb{R}}$ is a measure defined on a σ-algebra $\mathcal{X}$ of subsets of X. We say that f and g are *equal almost everywhere with respect to* μ (or *equal μ-almost everywhere*), denoted by

$$f = g \quad \mu\text{-a.e.,}$$

if $f(x) = g(x)$ for every $x \in X \backslash N$ for some $N \in \mathcal{X}$ such that $\mu(N) = 0$ (i.e., $f(x) = g(x)$ for every x except perhaps in a set of measure zero).

Example 2F. Let $\{f_n\}$ be a sequence of real-valued functions $f_n: X \to \mathbb{R}$ on a set X. We say that $\{f_n\}$ converges pointwise to a function $f: X \to \mathbb{R}$ if the real-valued sequence $\{f_n(x)\}$ converges to $f(x)$ for every $x \in X$. Thus pointwise convergence means convergence of $\{f_n(x)\}$ everywhere in X. Again, suppose $\mu: \mathcal{X} \to \overline{\mathbb{R}}$ is a measure defined on a σ-algebra $\mathcal{X}$ of subsets of X. We say that $\{f_n\}$ *converges to f almost everywhere with respect to* μ (or *converges μ-almost everywhere to f*), denoted by

$$f_n \to f \quad \mu\text{-a.e.} \qquad \text{or} \qquad \lim_n f_n = f \quad \mu\text{-a.e.,}$$

if $f_n(x) \to f(x)$ for every $x \in X \backslash N$ for some $N \in \mathcal{X}$ such that $\mu(N) = 0$ (i.e., $f_n(x) \to f(x)$ for every x except perhaps in a set of measure zero).

2.2   Signed Measure

Let $(X, \mathcal{X})$ be a measurable space. If μ and λ are two measures on $\mathcal{X}$ and α is nonnegative real number, then it is really easy to show that $\alpha\mu$ and $\mu + \lambda$ are again measures on $\mathcal{X}$ ("pointwise" defined: $(\alpha\mu)(E) = \alpha\mu(E)$ and $(\mu + \lambda)(E) = \mu(E) + \lambda(E)$ for every $E \in \mathcal{X}$). Therefore, it is readily verified that every (finite) linear combination $\sum_{i=1}^{n} \alpha_i \mu_i$ with nonnegative coefficients α_i of measures μ_i on $\mathcal{X}$ is again a measure on $\mathcal{X}$. The assumption of nonnegative coefficients is imposed to ensure nonnegativeness for the resulting measure (Definition 2.1(b)). If we ignore this requirement, then we may consider any real coefficients. In particular, we may consider the set function $\mu - \lambda$ on $\mathcal{X}$ defined by $(\mu - \lambda)(E) = \mu(E) - \lambda(E)$ for every $E \in \mathcal{X}$. However, if there exists an E in $\mathcal{X}$ such that $\mu(E) = \lambda(E) = +\infty$, then $\mu(E) - \lambda(E)$ is not well defined. An obvious way to avoid this problem is to assume that at least one of μ or λ does not take on the value $+\infty$. Another way consists in assuming that μ and λ are both real-valued measures.

Definition 2.3. Let $(X, \mathcal{X})$ be a measurable space. A real-valued function

$$\nu : \mathcal{X} \to \mathbb{R}$$

is a *signed measure* if it satisfies the following conditions.

(a) $\nu(\varnothing) = 0$.

(b) $\nu\left(\bigcup_n E_n\right) = \sum_n \nu(E_n)$

for every countable family $\{E_n\}$ of pairwise disjoint sets in $\mathcal{X}$ for which the series $\sum_n \nu(E_n)$ is unconditionally convergent.

In other words, a signed measure is a real-valued function on a σ-algebra that may fail to be a measure (a finite measure, that is) just because it may not satisfy the measure axiom of Definition 2.1(b). First observe that we are now dealing with real-valued functions, so the series in (b) must converge (otherwise the left-hand side of (b) would not be well defined). Moreover, recall that in Definition 2.1(c) we had a series of nonnegative terms, where convergence coincides with absolute convergence, which (in $\mathbb{R}$) means unconditional convergence. However, now we have a series of real numbers (so unconditional convergence is not a consequence of convergence) and the identity in (b) demands unconditional convergence (reason: the union in the left-hand side is order invariant, and so must be the series in the right-hand side) or, equivalently, absolute convergence.

The results in Proposition 2.2(b,c,d) still hold for a signed measure — essentially the same proof. Also note that any (finite) linear combination of signed measures is again a signed measure. If μ and λ are finite measures, then the function $\nu = \mu - \lambda$ is a signed measure. This was our very motivation for defining signed measures. There are other ways to get signed measures from measures (cf. Chapter 4). On the other hand, an interesting problem is to get measures from signed measure (cf. Chapter 7). The following example shows that what might seem the trivial way simply fails. The next two examples do exhibit measures generated by a signed measure.

Example 2G. Let $(X, \mathcal{X})$ be a measurable space. If $\nu : \mathcal{X} \to \mathbb{R}$ is a signed measure on $\mathcal{X}$, then the set function $\pi : \mathcal{X} \to \mathbb{R}$, defined for each $E \in \mathcal{X}$ by

$$\pi(E) = |\nu(E)|,$$

is *not* necessarily a measure. Indeed, if the signed measure ν is not a measure, then there is a set B in $\mathcal{X}$ for which $\nu(B) < 0$. If there exists a set A in $\mathcal{X}$ such that $A \cap B = \varnothing$ and $\nu(A) = -\nu(B)$, then $\nu(A \cup B) = 0$. Thus $A \subset A \cup B$ and $\pi(A) \not\leq \pi(A \cup B)$, so π is not a measure according to Proposition 2.2(a). Another way to say the same thing:

$$\pi(A \cup B) = |\nu(A \cup B)| = 0 < |\nu(A)| + |\nu(B)| = \pi(A) + \pi(B),$$

and therefore even finite additivity fails for π so that π is not a measure (cf. Definition 2.1(c) — a concrete example is exhibited in Problem 2.12).

Example 2H. Take a measurable space $(X, \mathcal{X})$ and, for each $\mathcal{X}$-measurable set E, consider the σ-algebra $\mathcal{E} = \wp(E) \cap \mathcal{X}$. Let $\nu \colon \mathcal{X} \to \mathbb{R}$ be a signed measure and consider the function $\mu \colon \mathcal{X} \to \mathbb{R}$ defined, for each $E \in \mathcal{X}$, by

$$\mu(E) = \sup_{A \in \mathcal{E}} \nu(A).$$

This is a measure on $\mathcal{X}$. Indeed, let $\{E_n\}$ be a arbitrary countable family of pairwise disjoint sets in $\mathcal{X}$, put $E = \bigcup E_n \in \mathcal{X}$, and take any $n \in \mathbb{N}$.

Claim. For every $\varepsilon > 0$ there exists an $\mathcal{X}$-measurable set $A_n \subseteq E_n$ such that

$$\nu(A_n) \leq \mu(E_n) \leq \nu(A_n) + \tfrac{\varepsilon}{2^n}.$$

Proof. First note that $\nu(A_n) \leq \mu(E_n)$ trivially by the very definition of μ. Now take an arbitrary $\varepsilon > 0$. If $\nu(A) + \varepsilon < \mu(E)$ for every $A \in \mathcal{E}$, then $\mu(E) + \varepsilon = \sup_{A \in \mathcal{E}} \nu(A) + \varepsilon < \mu(E)$, which is a contradiction. Thus, for every $\varepsilon > 0$ and every $E \in \mathcal{X}$ there is an $A_\varepsilon \in \mathcal{E}$ such that $\mu(E) \leq \nu(A_\varepsilon) + \varepsilon$. In particular, for arbitrary $n \in \mathbb{N}$ and $\varepsilon > 0$, put $\varepsilon_n = \tfrac{\varepsilon}{2^n}$ so that there exists an $\mathcal{X}$-measurable $A_n \subseteq E_n$ for which $\mu(E_n) \leq \nu(A_n) + \tfrac{\varepsilon}{2^n}$.   □

Therefore,

$$\sum_n \nu(A_n) \leq \sum_n \mu(E_n) \leq \sum_n \nu(A_n) + \varepsilon,$$

for $\sum_{n \in \mathbb{N}} \tfrac{1}{2^n} = 1$. Moreover, since $\{A_n\}$ is a disjoint sequence (because $\{E_n\}$ is) of sets in $\mathcal{X}$, and since ν is a signed measure, it follows that

$$\nu\left(\bigcup_n A_n\right) = \sum_n \nu(A_n).$$

Set $A = \bigcup_n A_n$ in $\mathcal{X}$ so that $A \subseteq E = \bigcup_n E_n$. Hence $A \in \mathcal{E}$ and

$$\nu(A) \leq \sum_n \mu(E_n) \leq \nu(A) + \varepsilon,$$

which implies that

$$\mu(E) \leq \sum_n \mu(E_n) \leq \mu(E) + \varepsilon$$

for every $\varepsilon > 0$ because $\mu(E) = \sup_{A \in \mathcal{E}} \nu(A)$. Thus, since $\bigcup_n E_n = E$,

$$\mu\left(\bigcup_n E_n\right) = \mu(E) = \sum_n \mu(E_n)$$

and the axiom (c) of Definition 2.2 (countable additivity) is satisfied. Since axioms (a) and (b) are trivially satisfied ($\wp(\varnothing) \cap \mathcal{X} = \{\varnothing\}$ and $\nu(\varnothing) = 0$, so $\mu(\varnothing) = 0$ and $\mu(E) \geq 0$ for every $E \in \mathcal{X}$), μ is a measure on $\mathcal{X}$.

Example 2I. Recall that a *covering* of a subset A of a set X is a collection of subsets of X that cover A (i.e., whose union includes A). Thus a partition of a set A is precisely a disjoint covering of it consisting of subsets of it. Let $\mathcal{X}$ be a σ-algebra of subsets of X and take any $\mathcal{X}$-measurable set E. A *measurable covering* of E is a covering of it consisting of $\mathcal{X}$-measurable sets. Consequently, a *measurable partition* of E is a disjoint covering of it made up of sets in $\mathcal{E} = \wp(E) \cap \mathcal{X}$. For each $n \in \mathbb{N}$ let $\boldsymbol{E}(n)$ be the collection of all measurable partitions of E containing n sets so that $\bigcup \boldsymbol{E}(n)$ is the collection of all *finite* measurable partitions of E. Let $\nu : \mathcal{X} \to \mathbb{R}$ be a signed measure and consider the function $\mu : \mathcal{X} \to \mathbb{R}$ defined, for each $E \in \mathcal{X}$, by

$$\mu(E) = \sup_{\{E_i\} \in \cup \boldsymbol{E}(n)} \sum_i |\nu(E_i)|.$$

Any finite partition $\{E_i\} \in \bigcup \boldsymbol{E}(n)$ can be written as $\{E_i\} = \{E_j\} \cup \{E_k\}$, where $\{E_j\}$ and $\{E_k\}$ are disjoint collections such that $\nu(E_j) \geq 0$ for each j and $\nu(E_k) \leq 0$ for each k. Put $E^+ = \bigcup_j E_j$ and $E^- = \bigcup_k E_k$ in $\mathcal{E}$ so that

$$\nu(E^+) = \nu\left(\bigcup_j E_j\right) = \sum_j \nu(E_j) \geq 0,$$

$$\nu(E^-) = \nu\left(\bigcup_k E_k\right) = \sum_k \nu(E_k) \leq 0,$$

and hence $\sum_i |\nu(E_i)| = \nu(E^+) - \nu(E^-)$, where $\{E^+, E^-\} \in \boldsymbol{E}(2)$. Thus

$$\mu(E) = \sup_{\{E^+, E^-\} \in \boldsymbol{E}(2)} \left(\nu(E^+) - \nu(E^-)\right) \quad \text{for every} \quad E \in \mathcal{X},$$

where the supremum is taken over all measurable partitions $\{E^+, E^-\}$ of E consisting of two sets such that $\nu(E^+) \geq 0$ and $\nu(E^-) \leq 0$. We shall see in Example 7A that μ is a measure on $\mathcal{X}$, which is called the *variation* of the signed measure ν. In fact, μ coincides with the "total variation" of ν, which is a measure on $\mathcal{X}$ that will be denoted by $|\nu|$ in Section 7.1.

2.3   Completion

A measure space is *complete* if the σ-algebra contains all subsets of sets of measure zero. In other words, $(X, \mathcal{X}, \mu)$ is a *complete measure space* if

$$N \in \mathcal{X}, \ \ \mu(N) = 0 \ \text{ and } \ A \subseteq N \quad \text{imply} \quad A \in \mathcal{X}.$$

We say that $\mathcal{X}$ is a *complete σ-algebra* (with respect to a measure μ) or that μ is a *complete measure* on $\mathcal{X}$ if (and only if) the measure space $(X, \mathcal{X}, \mu)$ is complete. Each measure space can be completed by adding up enough subsets of measure zero to the σ-algebra, as we shall see next. Let $(X, \mathcal{X}, \mu)$ be a measure space and consider the collection

$$\mathcal{N} = \{ N \in \mathcal{X} \colon \mu(N) = 0 \}$$

of all sets in $\mathcal{X}$ of measure zero. Let $\overline{\mathcal{X}}$ be the collection of all sets of the form $E \cup A$, where E is a set in $\mathcal{X}$ and A is a subset of some set in $\mathcal{N}$:

$$\overline{\mathcal{X}} = \{ \overline{E} \subseteq X \colon \overline{E} = E \cup A \text{ with } E \in \mathcal{X} \text{ and } A \subseteq N \text{ for some } N \in \mathcal{N} \}.$$

Define a function $\overline{\mu} \colon \overline{\mathcal{X}} \to \mathbb{R}$ as follows:

$$\overline{\mu}(\overline{E}) = \mu(E)$$

for every $\overline{E} \in \overline{\mathcal{X}}$, where E is any set in $\mathcal{X}$ for which $\overline{E} = E \cup A$ for some subset A of some N in $\mathcal{N}$. Note that this function $\overline{\mu}$ is well defined. Indeed, take an arbitrary $\overline{E} \in \overline{\mathcal{X}}$ and consider any pair of possible representations of it, namely, $\overline{E} = E_1 \cup A_1 = E_2 \cup A_2$, where E_1, E_2 are sets in $\mathcal{X}$ and A_1, A_2 are subsets of some sets N_1, N_2 in $\mathcal{N}$, respectively. Since $E_1 \subseteq E_1 \cup A_1 = E_2 \cup A_2 \subseteq E_2 \cup N_2$, and since $E_2 \cup N_2$ lies in $\mathcal{X}$ (Definition 1.1(c$'$)), we get from Proposition 2.2(a) and Definition 2.1(c) that $\mu(E_1) \leq \mu(E_2 \cup N_2) = \mu(E_2)$. Similarly, $\mu(E_2) \leq \mu(E_1 \cup N_1) = \mu(E_1)$. Thus $\mu(E_1) = \mu(E_2)$. This ensures that the function $\overline{\mu}$ is well defined: it assigns to each $\overline{E}$ the value $\mu(E)$, which is invariant for all representations of $\overline{E} = E \cup A$. Moreover,

(i) $\overline{\mathcal{X}}$ is a σ-algebra of subsets of X that includes the σ-algebra $\mathcal{X}$ of subsets of X (which means that $\mathcal{X}$ is a *sub-σ-algebra* of $\overline{\mathcal{X}}$), and

(ii) $\overline{\mu}$ is a measure on $\overline{\mathcal{X}}$ that agrees with μ on $\mathcal{X}$ (that is, that $\overline{\mu}$ is an *extension* of μ *over* $\overline{\mathcal{X}}$ or, equivalently, μ is a *restriction* of $\overline{\mu}$ to $\mathcal{X}$).

Proposition 2.4. *Let* $(X, \mathcal{X}, \mu)$ *be a measure space.*

(a) $\overline{\mathcal{X}}$ *is a σ-algebra of subsets of X such that* $\mathcal{X} \subseteq \overline{\mathcal{X}}$.

(b) $\overline{\mu} \colon \overline{\mathcal{X}} \to \mathbb{R}$ *is a measure on* $\overline{\mathcal{X}}$ *such that* $\overline{\mu}(E) = \mu(E)$ *for every* $E \in \mathcal{X}$.

(c) $(X, \overline{\mathcal{X}}, \overline{\mu})$ *is a complete measure space.*

Proof. To begin with, observe that $\mathcal{X} \subseteq \overline{\mathcal{X}}$ once $E = E \cup \varnothing$ for every $E \in \mathcal{X}$. In particular, the empty set and the whole set trivially lie in $\overline{\mathcal{X}}$. Take an arbitrary $\overline{E} \in \overline{\mathcal{X}}$. We shall show that $X \backslash \overline{E} \in \overline{\mathcal{X}}$. Indeed, if $\overline{E} = E \cup A$, then $X \backslash \overline{E} = X \backslash (E \cup A) = (X \backslash E) \cap (X \backslash A) = E' \backslash A$. Here $E' = X \backslash E$, which lies in $\mathcal{X}$ since $E \in \mathcal{X}$. Let N be any set in $\mathcal{N} \subseteq \mathcal{X}$ such that $A \subseteq N$. Thus $E' \backslash A = (E' \backslash N) \cup (N \backslash A) = E_1 \cup A_1$, where $E_1 = E' \backslash N$ lies in $\mathcal{X}$ because both E' and N lie in $\mathcal{X}$, and $A_1 = N \backslash A \subseteq N$. Therefore,

$$X \backslash \overline{E} = E' \backslash A = E_1 \cup A_1 \quad \text{lies in} \quad \overline{\mathcal{X}}.$$

Now take an arbitrary sequence $\{\overline{E}_n\}$ of sets in $\overline{\mathcal{X}}$ so that $\overline{E}_n = E_n \cup A_n$, where $E_n \in \mathcal{X}$ and $A_n \subseteq N_n$ with $N_n \in \mathcal{N}$, for each n. Put $E = \bigcup_n E_n$ in $\mathcal{X}$ (Definition 1.1.(c')), and $A = \bigcup_n A_n \subseteq N = \bigcup_n N_n$. Note that $N \in \mathcal{N}$ (Definition 2.1(c)). Thus $\overline{\mathcal{X}}$ is a σ-algebra (cf. Definition 1.1) because

$$\bigcup_n \overline{E}_n = \bigcup_n (E_n \cup A_n) = \left(\bigcup_n E_n \right) \cup \left(\bigcup_n A_n \right) = E \cup A \quad \text{lies in} \quad \overline{\mathcal{X}},$$

which completes the proof of (a). Observe that the function $\overline{\mu}$ agrees with the measure μ on $\mathcal{X}$ ($\overline{\mu}(E) = \mu(E)$ for every $E \in \mathcal{X}$) by the very definition of $\overline{\mu}$ (since $E = E \cup \varnothing$). In particular, $\overline{\mu}(\varnothing) = 0$ and $\overline{\mu}(\overline{E}) = \mu(E) \geq 0$ for every $\overline{E} \in \overline{\mathcal{X}}$. Moreover, suppose $\{\overline{E}_n\}$ is an arbitrary sequence of pairwise disjoint sets in $\overline{\mathcal{X}}$ so that $\overline{E}_n = E_n \cup A_n$, where $E_n \in \mathcal{X}$ and $A_n \subseteq N_n$ with $N_n \in \mathcal{N}$, for each n. Hence $\{E_n\}$ is a sequence of pairwise disjoint sets in $\mathcal{X}$, and therefore it follows from the above displayed identity that

$$\overline{\mu}\left(\bigcup_n \overline{E}_n \right) = \overline{\mu}(E \cup A) = \mu(E) = \mu\left(\bigcup_n E_n \right) = \sum_n \mu(E_n) = \sum_n \overline{\mu}(\overline{E}_n)$$

according to the definition of $\overline{\mu}$ since μ is a measure on $\mathcal{X}$ (Definition 2.1(c)). Thus $\overline{\mu}$ is a measure on $\overline{\mathcal{X}}$ (cf. Definition 2.1). This proves (b). Finally, put

$$\overline{\mathcal{N}} = \{ \overline{N} \in \overline{\mathcal{X}} : \overline{\mu}(\overline{N}) = 0 \}.$$

If $\overline{N} \in \overline{\mathcal{N}}$, then $\overline{N} \in \overline{\mathcal{X}}$. Thus the set $\overline{N}$ is of the form $\overline{N} = N' \cup A$ with $N' \in \mathcal{X}$ and $A \subseteq N$ for some $N \in \mathcal{N}$ (according to the definition of $\overline{\mathcal{X}}$). But $\mu(N') = \overline{\mu}(\overline{N}) = 0$ (cf. definition of $\overline{\mu}$), and hence $N' \in \mathcal{N}$. Outcome:

$$\overline{N} \in \overline{\mathcal{N}} \quad \text{implies} \quad \overline{N} = N' \cup A \text{ with } A \subseteq N \text{ for some } N, N' \in \mathcal{N}. \quad (*)$$

Moreover, if $A \subseteq N \in \mathcal{N} \subseteq \mathcal{X}$, then $A = \varnothing \cup A$ must lie in $\overline{\mathcal{X}}$ since $\varnothing \in \mathcal{X}$:

$$A \subseteq N \in \mathcal{N} \quad \text{implies} \quad A \in \overline{\mathcal{X}}. \quad (**)$$

Therefore, if $\overline{A}$ is any subset of any set $\overline{N}$ in $\overline{\mathcal{N}}$, then $\overline{A} \subseteq \overline{N} = N' \cup A$ with $A \subseteq N$ for some $N, N' \in \mathcal{N}$ by $(*)$. Thus $\overline{A} = A' \cup A''$, where $A' \subseteq N'$ and

$A'' \subseteq A \subseteq N$. However, both A' and A'' lie in $\overline{\mathcal{X}}$ by $(\ast\ast)$, and hence $\overline{A} \in \overline{\mathcal{X}}$ (Definition 1.1(c')). That is, $\overline{\mathcal{X}}$ is a complete σ-algebra with respect to the measure $\overline{\mu}$, which completes the proof of (c). $\qquad\qquad\square$

We refer to $(X, \overline{\mathcal{X}}, \overline{\mu})$ as the *completion of the measure space* $(X, \mathcal{X}, \mu)$. Accordingly, we say that $\overline{\mathcal{X}}$ is the *completion of the σ-algebra* $\mathcal{X}$ (with respect to the measure μ) and $\overline{\mu}$ is the *completion of the measure* μ on $\mathcal{X}$.

Remark: The Lebesgue measure λ on the σ-algebra $\mathfrak{R}$ of Borel sets as in Example 2C does not make a complete measure space $(\mathbb{R}, \mathfrak{R}, \lambda)$. That is, on the Borel algebra $\mathfrak{R}$ the Lebesgue measure λ is not complete. However, we shall see in Chapter 8 how to build its completion, where the Lebesgue measure $\overline{\lambda}$ on the *Lebesgue algebra* $\overline{\mathfrak{R}}$ (the completion of the Borel algebra $\mathfrak{R}$ with respect to λ) makes a complete measure space $(\mathbb{R}, \overline{\mathfrak{R}}, \overline{\lambda})$. Warning: the notation $\overline{\mathfrak{R}}$ is tricky; it does not mean the σ-algebra generated by the collection of all open intervals of the extended real line $\overline{\mathbb{R}}$.

2.4 Problems

Problem 2.1. Let $(X, \mathcal{X})$ be a measurable space and let $\mu \colon \mathcal{X} \to \overline{\mathbb{R}}$ be a measure on $\mathcal{X}$.

(a) Show that μ is a finite measure if and only if there exists $0 \le \alpha \in \mathbb{R}$ such that $\mu(E) < \alpha$ for all $E \in \mathcal{X}$.

Indeed, it is readily verified by Proposition 2.2(a) that μ is finite if and only if $\mu(X) < \infty$. Now consider the definition of a σ-finite measure and assume further that there exists an $\mathcal{X}$-valued sequence $\{E_n\}$ and a real (nonnegative) number α such that $\mu(E_n) \le \alpha$ for all n and $X = \bigcup_n E_n$. If this is the case, then we say that μ is *uniformly σ-finite*.

(b) Check that the counting measure of Example 2B is uniformly σ-finite.

(c) Check that the Lebesgue measure of Example 2C is uniformly σ-finite.

(d) When (i.e., for which class of f) is the Borel–Stieltjes measure λ_f of Example 2D finite? When is it uniformly σ-finite?

Problem 2.2. It is clear that every finite measure is uniformly σ-finite and that every uniformly σ-finite measure is σ-finite. However, the converses fail. Indeed, consider the σ-algebra $\wp(\mathbb{N})$ of all subsets of the natural numbers $\mathbb{N}$. We saw in the previous problem that the counting measure on $\wp(\mathbb{N})$ is uniformly σ-finite, and it is obviously not finite (the value of it at $\mathbb{N}$ is not finite). Now consider the function $\mu \colon \wp(\mathbb{N}) \to \overline{\mathbb{R}}$ defined by

$$\mu(E) = \sum_{k \in E} k$$

for every $E \in \mathcal{P}(\mathbb{N})$ (by convention, the empty sum is null). Show that this is a measure on $\mathcal{P}(\mathbb{N})$ that is σ-finite but not uniformly σ-finite.

Problem 2.3. Show that a measure $\mu \colon \mathcal{X} \to \overline{\mathbb{R}}$ on a σ-algebra $\mathcal{X}$ of subsets of a set X is σ-finite if and only if there exists a countable family $\{E_k\}$ of *disjoint* sets in $\mathcal{X}$ such that $\mu(E_k) < \infty$ for every k and $X = \bigcup_k E_k$.

Hint: Every sequence of sets $\{X_n\}_{n \in \mathbb{N}}$ has a *disjointification* $\{Y_n\}_{n \in \mathbb{N}}$ (i.e., $\{Y_n\}_{n \in \mathbb{N}}$ is a sequence of pairwise disjoint sets and $\bigcup_n Y_n = \bigcup_n X_n$).

Problem 2.4. Take any set X and let $\mathcal{X}$ be a σ-algebra of subsets of X. A measure $\mu \colon \mathcal{X} \to \overline{\mathbb{R}}$ is called *semifinite* if every measurable set of infinite measure includes a measurable set of arbitrarily large finite measure.

(a) Show that every σ-finite measure is semifinite.

   Hint: Suppose a measure is not semifinite but σ-finite (Problem 2.3).

Now assume that X is an uncountable set. Let $\#$ stand for cardinality and consider the functions $\mu \colon \mathcal{X} \to \overline{\mathbb{R}}$ and $\nu \colon \mathcal{X} \to \overline{\mathbb{R}}$ defined, for each $E \in \mathcal{X}$, by

$$\mu(E) = \begin{cases} \#(E), & E \text{ is finite,} \\ +\infty, & E \text{ is infinite,} \end{cases} \quad \text{and} \quad \nu(E) = \begin{cases} 0, & E \text{ is countable,} \\ +\infty, & E \text{ is uncountable.} \end{cases}$$

(b) Show that μ is a measure on $\mathcal{X}$ (the counting measure on an *uncountable* set) that is semifinite but not σ-finite.

(c) Show that ν is a measure on $\mathcal{X}$ that is not semifinite (thus not σ-finite according to (a)).

Hint: Every infinite set has a countably infinite subset. A countable union of countable sets is countable.

Problem 2.5. Let $\mathcal{X}$ be a σ-algebra of subsets of an infinite set X for which all singletons are measurable sets and consider the set functions $\pi \colon \mathcal{X} \to \overline{\mathbb{R}}$ and $\rho \colon \mathcal{X} \to \overline{\mathbb{R}}$ defined, for each $E \in \mathcal{X}$, by

$$\pi(E) = \begin{cases} 0, & E \text{ is finite,} \\ +\infty, & E \text{ is infinite,} \end{cases} \quad \text{and} \quad \rho(E) = \begin{cases} 0, & E \text{ is finite,} \\ 1, & E \text{ is infinite.} \end{cases}$$

Why are these functions not measures?

Problem 2.6. Let X be an uncountable set and let $\mathcal{X}$ be a collection of those subsets of X that either are countable or are the complement of a countable subset of X. Show that $\mathcal{X}$ is a σ-algebra of subsets of X and consider the function $\mu\colon \mathcal{X} \to \mathbb{R}$ defined, for each $E \in \mathcal{X}$, by

$$\mu(E) = \begin{cases} 0, & E \text{ is countable,} \\ 1, & X\backslash E \text{ is countable.} \end{cases}$$

Show that μ is a measure on $\mathcal{X}$, a probability measure, that is.

Problem 2.7. Let $\lambda\colon \mathfrak{R} \to \mathbb{R}$ be the Lebesgue measure on the Borel algebra $\mathfrak{R}$ of subsets of $\mathbb{R}$ (cf. Example 2C). Prove the following propositions.

(a) Every singleton of $\mathbb{R}$ is $\mathfrak{R}$-measurable and has measure zero.

(b) Every countable subset of $\mathbb{R}$ is $\mathfrak{R}$-measurable and has measure zero.

(c) If $\alpha < \beta$, then the intervals (α, β), $[\alpha, \beta)$, $(\alpha, \beta]$, $[\alpha, \beta]$ are $\mathfrak{R}$-measurable and $\lambda((\alpha, \beta)) = \lambda([\alpha, \beta)) = \lambda((\alpha, \beta]) = \lambda([\alpha, \beta]) = \beta - \alpha$.

(d) Every nonempty open subset U of $\mathbb{R}$ is $\mathfrak{R}$-measurable and $\lambda(U) > 0$. (*Hint:* $\mathbb{R}$ has a countable base of open intervals — see Problem 1.14.)

(e) Every bounded $\mathfrak{R}$-measurable subset of $\mathbb{R}$ has a finite measure.

(f) A closed and bounded subset K of $\mathbb{R}$ is $\mathfrak{R}$-measurable and $\lambda(K) < \infty$.

Show that λ is uniformly σ-finite by exhibiting a countably infinite family $\{E_k\}$ of *disjoint sets* in $\mathcal{X}$ such that $\mu(E_k) = 1$ for all k and $X = \bigcup_k E_k$.

Problem 2.8. Let $\{E_n\}$ be a sequence of sets in a σ-algebra $\mathcal{X}$ and let $\mu\colon \mathcal{X} \to \overline{\mathbb{R}}$ be any measure on $\mathcal{X}$. Apply Proposition 2.2 to show that

(a) $\mu\left(\bigcup_n E_n\right) = \lim_n \mu\left(\bigcup_{i=1}^n E_i\right)$,

(b) $\mu\left(\bigcup_n E_n\right) \leq \sum_n \mu(E_n)$.

Hint: Set $A_n = \bigcup_{i=1}^n E_i$ to prove (a) and $B_{n+1} = E_{n+1}\backslash\left(\bigcup_{i=1}^n E_i\right)$ so that $B_n \subseteq E_n$ and $\{B_n\}$ is pairwise disjoint to prove (b).

Problem 2.9. The *Cantor set* is a rather well-known subset of the interval $[0, 1]$ of the real line $\mathbb{R}$. The reader is referred to the bibliography mentioned in the Suggested Reading section for many aspects of it. Roughly speaking, the Cantor set C is the intersection of a decreasing sequence $\{C_n\}$ of closed subsets of $C_0 = [0, 1]$ obtained by successive removal of the central open third. Among the main properties of the Cantor set $C = \bigcap_n C_n \subset \mathbb{R}$ we have: C is nonempty, closed, and bounded; it has an empty interior and has no isolated point; it is uncountable and totally disconnected. Consider the Lebesgue measure $\lambda\colon \mathfrak{R} \to \mathbb{R}$ as in Example 2C (also see Problem 2.7) and show that the Cantor set has measure zero. That is, C lies in $\Re$ and $\lambda(C) = 0$. (*Hint:* Each C_n consists of 2^n disjoint intervals of length $\frac{1}{3^n}$.)

Problem 2.10. Consider the setup and the construction of the previous problem, where each set C_n (for $n \in \mathbb{N}$) is obtained from C_{n-1} by removing 2^{n-1} central open subintervals, each of length $\frac{1}{3^n}$. Now, instead of removing at each iteration 2^{n-1} central open subintervals of length $\frac{1}{3^n}$, remove at each iteration 2^{n-1} central open subintervals of length $\frac{1}{4^n}$. Let $\{S_n\}$ be the resulting decreasing sequence of closed subsets of the unit interval $S_0 = [0,1]$. Show that the length of each S_n is $\lambda(S_n) = \frac{1}{2} + \frac{1}{2^{n+1}}$ and conclude that $S = \bigcap_n S_n \subset \mathbb{R}$ lies in $\Re$ and $\lambda(S) = \frac{1}{2}$. (*Hint:* Proposition 2.2(c).)

Problem 2.11. Let $\mu: \mathcal{X} \to \overline{\mathbb{R}}$ be a measure on a σ-algebra $\mathcal{X}$ of subsets of a set X. Take an arbitrary $\mathcal{X}$-measurable set A and consider the σ-algebra $\mathcal{A} = \mathscr{P}(A) \cap \mathcal{X}$ of subsets of A. It is plain that $\mathcal{A} \subseteq \mathcal{X}$ ($\mathcal{A}$ is a σ-algebra of subsets of $A \in \mathcal{X}$, which is a subcollection of the σ-algebra $\mathcal{X}$ of subsets of X). Define the functions $\mu_A: \mathcal{X} \to \overline{\mathbb{R}}$ and $\mu|_A: \mathcal{A} \to \overline{\mathbb{R}}$ by

$$\mu_A(E) = \mu(E \cap A) \quad \text{for every} \quad E \in \mathcal{X},$$

$$\mu|_A(E) = \mu(E) \quad \text{for every} \quad E \in \mathcal{A}.$$

Show that $\mu_A: \mathcal{X} \to \overline{\mathbb{R}}$ and $\mu|_A: \mathcal{A} \to \overline{\mathbb{R}}$ are measures on $\mathcal{X}$ and on $\mathcal{A}$, respectively. (*Hint:* $\left(\bigcup_n E_n\right) \cap A = \bigcup_n (E_n \cap A)$.) The measure $\mu|_A$ is *the restriction of both μ and μ_A to $\mathcal{A}$, so μ and μ_A are (different) extensions of $\mu|_A$ over $\mathcal{X}$* — all these measures coincide on $\mathcal{A}$.

Problem 2.12. Let $\lambda: \Re \to \overline{\mathbb{R}}$ be the Lebesgue measure on the Borel algebra $\Re$ of subsets of the real line $\mathbb{R}$ (cf. Example 2C and Problem 2.7). Set $A = [1,2]$, $B = [-2,-1]$, and consider the measures $\lambda_A: \Re \to \overline{\mathbb{R}}$ and $\lambda_B: \Re \to \overline{\mathbb{R}}$ defined in the previous problem. Verify that both λ_A and λ_B are finite measures so that $\nu = \lambda_A - \lambda_B: \Re \to \mathbb{R}$ is a signed measure,

$$\nu(A \cup B) = \lambda_A(A \cup B) - \lambda_B(A \cup B) = 0,$$

$$|\nu(A)| + |\nu(B)| = |\lambda_A(A) - \lambda_B(A)| + |\lambda_A(B) - \lambda_B(B)| = 2.$$

Thus conclude that the set function $\pi: \Re \to \mathbb{R}$, defined for each $E \in \Re$ by $\pi(E) = |\nu(E)| = |\lambda_A(E) - \lambda_B(E)|$, is not a measure (see Example 2G).

Problem 2.13. Let $\mu: \Re \to \overline{\mathbb{R}}$ be a measure on the Borel algebra $\Re$ of subsets of $\mathbb{R}$ such that $\mu(K) < \infty$ for every closed and bounded (Borel) subset K of $\mathbb{R}$. This is called a *Borel measure*. Show that the Lebesgue measure of Example 2C is a Borel measure and that every Borel measure is σ-finite.

If μ is a Borel measure, then its *support* is the set $[\mu] = \mathbb{R}\backslash U$, where U is the union of all open (Borel) sets of measure zero. Show that $[\mu]$ is a closed (Borel) set and $\mathbb{R}\backslash[\mu]$ is the largest (inclusion ordering) open (Borel) set of measure zero. Show that a point $\alpha \in \mathbb{R}$ is not in the support of μ if and only if there exists an open subset of measure zero that contains α. Let A be a closed (Borel) set with $0 < \mu(A) < \infty$, take the σ-algebra $\mathcal{A} = \wp(A) \cap \mathfrak{R}$, and consider the restriction $\nu = \mu|_A : \mathcal{A} \to \mathbb{R}$ of μ to $\mathcal{A}$, which is a finite measure. Samples: if μ is a finite measure, then A may be any closed subset of $\mathbb{R}$ of nonzero measure (e.g., $A = \mathbb{R}$ and $\mathcal{A} = \mathfrak{R}$); if μ is the Lebesgue measure, then A may be a closed and bounded *nondegenerate* interval (i.e., one that is not a singleton). Show that the support $[\nu]$ of ν is the smallest (inclusion ordering) closed (Borel) subset of A such that $\nu([\nu]) = \nu(A)$.

Problem 2.14. Let $(X, \mathcal{X})$ be a measurable space. Show that the sum of σ-finite measures on $\mathcal{X}$ is again a σ-finite measure on $\mathcal{X}$.

Hint: Let $\{E_n\}$ and $\{F_n\}$ in $\mathcal{X}$ be such that $\mu(E_n) < \infty$, $\lambda(F_n) < \infty$, and $\bigcup_n E_n = \bigcup_n F_n = X$. If $\lambda(E_i) < \infty$, then take E_i. If $\lambda(E_k) = \infty$, then take $\{F_{n_j}\}$ such that $E_k = \bigcup_j F_{n_j}$. Show that the collection of all those E_i and $\{F_{n_j}\}$ consists of a countable collection that makes $(\mu + \lambda)$ σ-finite.

Problem 2.15. Let $(X, \overline{\mathcal{X}}, \overline{\mu})$ be a completion of a measure space $(X, \mathcal{X}, \mu)$. Show that if $\overline{f} : X \to \overline{\mathbb{R}}$ is an $\overline{\mathcal{X}}$-measurable function, then there exists an $\mathcal{X}$-measurable function $f : X \to \overline{\mathbb{R}}$ such that $f = \overline{f}$ μ-a.e.

Hint: Take any $q \in \mathbb{Q}$, put $\overline{E}_q = \overline{f}^{-1}((q, \infty))$ in $\overline{\mathcal{X}}$, write $\overline{E}_q = E_q \cup A_q$ with $E_q \in \mathcal{X}$ and $A_q \subseteq N_q \in \mathcal{N}$, set $N = \bigcup N_q$, show that $N \in \mathcal{N}$ (countable union) so that $A_q \subseteq N$, and put $f(x) = \overline{f}(x)$ if $x \in X\backslash N$ and $f(x) = 0$ if $x \in N$. Verify that $f^{-1}((q, \infty))$ is either $\overline{E}_q\backslash N$ or $\overline{E}_q \cup N$, and hence $f^{-1}((q, \infty))$ lies in $\mathcal{X}$ all $q \in \mathbb{Q}$ so that f is $\mathcal{X}$-measurable (Problem 1.1).

Problem 2.16. Let $(X, \mathcal{X})$ be a measurable space. A complex-valued set function $\nu : \mathcal{X} \to \mathbb{C}$ satisfying axioms (a) and (b) of Definition 2.3, with absolute convergence on the right-hand side of (b), is a *complex measure*. Show that every complex measure ν on $\mathcal{X}$ can be expressed as $\nu = \nu_1 + i\nu_2$, where ν_1 and ν_2 are (real-valued) signed measures on $\mathcal{X}$.

Suggested Reading

Bartle [3], Berberian [6], Halmos [13], Royden [22]. For construction and properties of the Cantor set, the reader is referred to [4], [8], [17], [20].

3

Integral

3.1 Nonnegative Functions

A *simple function* is just a real-valued function $\varphi\colon X \to \mathbb{R}$ on any set X that has a finite range (i.e., a function that takes on only a finite number of distinct values). It is readily verified that φ is a simple function if and only if it can be represented as a linear combination of characteristic functions,

$$\varphi = \sum_{i=1}^{n} \alpha_i \chi_{E_i},$$

where $\{E_i\}_{i=1}^{n}$ is a finite collection of subsets of X (i.e., $\{E_i\}_{i=1}^{n} \subseteq \wp(X)$) and $\{\alpha_i\}_{i=1}^{n}$ is a finite (similarly indexed) set of real numbers. The foregoing expression is referred to as a *representation* of the simple function φ, which is not unique. However, there exists a unique representation of φ such that $\{\alpha_i\}_{i=1}^{n}$ is a set of *distinct* coefficients and $\{E_i\}_{i=1}^{n}$ is a *partition* of X (i.e., a collection of *disjoint* sets that *cover* X). Indeed, this is the representation of φ for which the set $\{\alpha_i\}_{i=1}^{n}$ is precisely the range of φ and, for each index i, $E_i = \{x \in X\colon \varphi(x) = \alpha_i\}$. Such a unique representation is called the *canonical* (or *standard*) *representation* of φ.

Let $\mathcal{X}$ be an arbitrary σ-algebra of subsets of X. Since a characteristic function $\chi_E\colon X \to \{0,1\}$ is a measurable function if and only if E is a measurable set (cf. Example 1B), it follows that a simple function $\varphi\colon X \to \mathbb{R}$ is a measurable function if and only if $\{E_i\}_{i=1}^{n}$ is a collection of measurable sets (i.e., $\{E_i\}_{i=1}^{n} \subseteq \mathcal{X}$) for any representation of it (cf. Proposition 1.5). We shall deal with measurable simple functions only. It is a simple matter to show that sum and (real) scalar multiple of measurable simple functions are again measurable simple functions, so the collection of all $\mathcal{X}$-measurable simple functions forms a linear subspace of the real linear space of all $\mathcal{X}$-measurable functions (and hence a linear space itself — cf. remarks that follow the proof of Proposition 1.5). In fact, we shall be dealing with non-negative measurable simple functions φ, which means that the coefficients $\{\alpha_i\}_{i=1}^n$ of any representation of φ is a set of nonnegative numbers.

Definition 3.1. Let $(X, \mathcal{X}, \mu)$ be a measure space. Take a nonnegative measurable simple function $\varphi \colon X \to \mathbb{R}$ and consider any representation of it,

$$\varphi = \sum_{i=1}^n \alpha_i \chi_{E_i}.$$

The *integral* of φ with respect to μ is the extended real number

$$\int \varphi \, d\mu = \sum_{i=1}^n \alpha_i \mu(E_i).$$

In particular, $\int \chi_E \, d\mu = \mu(E)$ for every $E \in \mathcal{X}$. It is easily seen that the integral of a nonnegative measurable simple function φ is independent of its representation. Therefore, the concept of integral of a simple function is unambiguously defined and we may assume the canonical representation of φ without loss of generality. To force the integral of the null function ($\varphi = 0$) to be well defined and equal to zero for every measure, including nonfinite measures (i.e., if $\mu(X) = +\infty$), we declare that $0 \cdot +\infty = 0$. The next proposition synthesizes the basic properties of the integral of simple functions. The first two properties ensure that the integral (as a functional of the collection of all nonnegative measurable simple functions into $\overline{\mathbb{R}}$) is nonnegative homogeneous and additive. The third property shows how the integral of a simple function with respect to a measure yields a new measure. The reader is invited to prove this proposition in Problem 3.2.

Proposition 3.2. Let $(X, \mathcal{X}, \mu)$ be a measure space. If φ and ψ are non-negative measurable simple functions and γ is a nonnegative real number, then $\gamma\varphi$ and $\varphi + \psi$ are nonnegative measurable simple functions and

(a) $\int \gamma\varphi \, d\mu = \gamma \int \varphi \, d\mu$,

(b) $\int (\varphi + \psi) \, d\mu = \int \varphi \, d\mu + \int \psi \, d\mu$,

(c) $\lambda(E) = \int \varphi \chi_E \, d\mu$ for every $E \in \mathcal{X}$ defines a measure $\lambda \colon \mathcal{X} \to \overline{\mathbb{R}}$.

Let $(X, \mathcal{X})$ be a measurable space and recall the notation $\mathcal{M}(X, \mathcal{X})$, or simply $\mathcal{M}$, for the collection of all $\mathcal{X}$-measurable functions (cf. Chapter 1).

Now let $\mathcal{M}(X, \mathcal{X})^+$, or simply $\mathcal{M}^+$ if the measurable space is clear in the context, be the collection of all nonnegative functions from $\mathcal{M}(X, \mathcal{X})$:

$$\mathcal{M}^+ = \mathcal{M}(X, \mathcal{X})^+ = \big\{ f: X \to \overline{\mathbb{R}}: \ f \text{ is } \mathcal{X}\text{-measurable and } f(x) \geq 0 \ \forall x \in X \big\}.$$

Note that extended real-valued functions are allowed in $\mathcal{M}$ and $\mathcal{M}^+$ but, of course, these collections also contain the real-valued functions. In particular, nonnegative $\mathcal{X}$-measurable simple functions lie in $\mathcal{M}(X, \mathcal{X})^+$. Given an arbitrary function f in $\mathcal{M}(X, \mathcal{X})^+$, consider the set Φ_f^+ of all simple functions φ in $\mathcal{M}(X, \mathcal{X})^+$ that are dominated by f,

$$\Phi_f^+ = \big\{ \varphi \in \mathcal{M}^+: \ \varphi \text{ is simple and } 0 \leq \varphi(x) \leq f(x) \ \forall x \in X \big\}.$$

Definition 3.3. Let $(X, \mathcal{X}, \mu)$ be a measure space. The *integral* of a function $f \in \mathcal{M}(X, \mathcal{X})^+$ with respect to μ is the extended real number

$$\int f \, d\mu = \sup_{\varphi \in \Phi_f^+} \int \varphi \, d\mu.$$

Moreover, the *integral* of f *over* a measurable set E with respect to μ is

$$\int_E f \, d\mu = \int f \chi_E \, d\mu \quad \text{in} \quad \overline{\mathbb{R}},$$

which is well defined since $\chi_E \in \mathcal{M}(X, \mathcal{X})^+$ (Example 1B), and so $f \chi_E$ lies in $\mathcal{M}(X, \mathcal{X})^+$ (Proposition 1.9), for every $E \in \mathcal{X}$. This defines a set function $\lambda: \mathcal{X} \to \overline{\mathbb{R}}$ given, for each set $E \in \mathcal{X}$, by $\lambda(E) = \int_E f \, d\mu$.

3.2   The Monotone Convergence Theorem

The context for this section is a fixed measure space $(X, \mathcal{X}, \mu)$. Our first aim is to extend the results of Proposition 3.2 from nonnegative simple functions to arbitrary nonnegative functions, which will come as a first consequence of a pivotal convergence theorem due to Beppo Levi. This is Theorem 3.4 below, one of the fundamental results in the theory of integration. Recall that a sequence $\{f_n\}$ of functions $f_n: X \to \overline{\mathbb{R}}$ is *increasing* if $f_n \leq f_{n+1}$ (i.e., $f_n(x) \leq f_{n+1}(x)$ for every $x \in X$) and *decreasing* if $f_{n+1} \leq f_n$ (i.e., $f_{n+1}(x) \leq f_n(x)$ for every $x \in X$) for each n. If it is either increasing or decreasing, then we say that the sequence is *monotone*. A monotone sequence of *extended* real-valued functions converges pointwise. (Why?)

Theorem 3.4. (Monotone Convergence Theorem). *Take a measure space* $(X, \mathcal{X}, \mu)$. *If* $\{f_n\}$ *is an increasing sequence of functions in* $\mathcal{M}(X, \mathcal{X})^+$, *converging pointwise to a function* $f: X \to \overline{\mathbb{R}}$, *then* $f \in \mathcal{M}(X, \mathcal{X})^+$ *and*

$$\int f \, d\mu = \lim_n \int f_n \, d\mu.$$

Proof. Take an arbitrary $x \in X$. Since $f_n(x) \geq 0$ for each n and $f(x) = \lim_n f_n(x)$, it follows that $f(x) \geq 0$, and therefore $f \in \mathcal{M}(X, \mathcal{X})^+$ according to Proposition 1.8. Since $f_n \leq f_{n+1}$ and $f = \lim_n f_n$, we get $f_n \leq f_{n+1} \leq f$, and so $\int f_n \, d\mu \leq \int f_{n+1} \, d\mu \leq \int f \, d\mu$, for every n, which implies that the extended real-valued increasing sequence $\{\int f_n \, d\mu\}$ converges and

$$\lim_n \int f_n \, d\mu \leq \int f \, d\mu. \tag{$*$}$$

On the other hand, take an arbitrary simple function φ in $\mathcal{M}(X, \mathcal{X})^+$ such that $0 \leq \varphi \leq f$ (i.e., an arbitrary $\varphi \in \Phi_f^+$). Let α be any real number in $(0, 1)$ and set $\psi = \alpha\varphi$, a simple function in $\mathcal{M}(X, \mathcal{X})^+$ with the property that $\psi(x) = 0$ if $f(x) = 0$ and $0 \leq \psi(x) < f(x)$ if $f(x) \neq 0$. For each n put

$$E_n = \{x \in X \colon \psi(x) \leq f_n(x)\} = \{x \in X \colon f_n(x) - \psi(x) \geq 0\}.$$

Since both f_n and ψ are measurable functions, $f_n - \psi$ is measurable, and so each E_n is a measurable set (Proposition 1.9). Hence, for each n,

$$\alpha \int_{E_n} \varphi \, d\mu = \int_{E_n} \psi \, d\mu \leq \int_{E_n} f_n \, d\mu \leq \int f_n \, d\mu$$

(cf. Proposition 3.2(a,c)). Since $\{f_n\}$ is increasing, $\{E_n\}$ is increasing as well. Moreover, since $f_n \nearrow f$ (i.e., $\{f_n\}$ is increasing and converges to f) and $0 \leq \psi(x) < f(x)$ if $f(x) \neq 0$, it follows that for each $x \in X$ there exists an m for which $\psi(x) \leq f_m(x) \leq f(x)$, and so x lies in E_m. This implies that $X = \bigcup_n E_n$. Thus $\{E_n\}$ is an increasing sequence of sets in $\mathcal{X}$ that cover X. Then, by Propositions 3.2(c) and 2.2(c),

$$\int \varphi \, d\mu = \lambda(X) = \lim_n \lambda(E_n) = \lim_n \int_{E_n} \varphi \, d\mu.$$

Therefore, by the above displayed expressions,

$$\alpha \int \varphi \, d\mu \leq \lim_n \int f_n \, d\mu,$$

and hence

$$\int \varphi \, d\mu = \sup_{\alpha \in (0,1)} \alpha \int \varphi \, d\mu \leq \lim_n \int f_n \, d\mu$$

so that

$$\int f \, d\mu = \sup_{\varphi \in \Phi_f^+} \int \varphi \, d\mu \leq \lim_n \int f_n \, d\mu. \tag{$**$}$$

The inequalities $(*)$ and $(**)$ imply $\int f \, d\mu = \lim_n \int f_n \, d\mu$. $\square$

The Monotone Convergence Theorem gave us the very first symptom of continuity for the integral transformation, and it will also give us the first symptoms of linearity (Proposition 3.5(a,b)). Of course, all this will make sense only when the domain of the integral transformation is equipped with the proper algebraic and topological structures, what we shall do in Chapter 5. Theorem 3.4 also shows how the integral of a *nonnegative* function with respect to a measure yields a new measure (Proposition 3.5(c)).

Proposition 3.5. *Let $(X, \mathcal{X}, \mu)$ be a measure space. If f and g are functions in $\mathcal{M}(X, \mathcal{X})^+$ and γ is a nonnegative real number, then γf and $f + g$ lie in $\mathcal{M}(X, \mathcal{X})^+$ and*

(a) $\int \gamma f \, d\mu = \gamma \int f \, d\mu$,

(b) $\int (f + g) \, d\mu = \int f \, d\mu + \int g \, d\mu$,

(c) $\lambda(E) = \int_E f \, d\mu$ *for every $E \in \mathcal{X}$ defines a measure $\lambda \colon \mathcal{X} \to \overline{\mathbb{R}}$.*

Proof. Since f and g lie in $\mathcal{M}(X, \mathcal{X})^+$, it follows by Proposition 1.9 that γf and $f + g$ lie in $\mathcal{M}(X, \mathcal{X})^+$ and, by Problem 1.6, that there exist increasing sequences $\{\varphi_n\}$ and $\{\psi_n\}$ of simple functions in $\mathcal{M}(X, \mathcal{X})^+$ such that $f = \lim_n \varphi_n$ and $g = \lim_n \psi_n$.

(a) If $\gamma \geq 0$, then $\{\gamma \varphi_n\}$ is an increasing sequence of simple functions in $\mathcal{M}(X, \mathcal{X})^+$ that converges to γf. Therefore, Proposition 3.2(a) and Theorem 3.4 (the Monotone Convergence Theorem) ensure that

$$\int \gamma f \, d\mu = \lim_n \int \gamma \varphi_n \, d\mu = \gamma \lim_n \int \varphi_n \, d\mu = \gamma \int f \, d\mu.$$

(b) Observe that $\{\varphi_n + \psi_n\}$ is an increasing sequence of simple functions in $\mathcal{M}(X, \mathcal{X})^+$ that converges to $f + g$. Thus, again, Proposition 3.2(b) and the Monotone Convergence Theorem ensure that

$$\int (f + g) \, d\mu = \lim_n \int (\varphi_n + \psi_n) \, d\mu = \lim_n \left(\int \varphi_n \, d\mu + \int \psi_n \, d\mu \right)$$

$$= \lim_n \int \varphi_n \, d\mu + \lim_n \int \psi_n \, d\mu = \int f \, d\mu + \int g \, d\mu.$$

(c) $\lambda(\varnothing) = 0$ because $\chi_\varnothing = 0$ so that $\int_\varnothing f \, d\mu = 0$, and $\lambda(E) \geq 0$ for all E in $\mathcal{X}$ by the definition of the integral of f in $\mathcal{M}(X, \mathcal{X})^+$. To verify countable additivity (Definition 2.1(c)), take any sequence $\{E_n\}$ of pairwise disjoint sets in $\mathcal{X}$. Since $\sum_{n=1}^m f \chi_{E_n} = f \chi_{\bigcup_{n=1}^m E_n}$ put, for each integer $m \geq 1$,

$$f_m = \sum_{n=1}^m f \chi_{E_n} = f \chi_{\bigcup_{n=1}^m E_n}.$$

It is readily verified that $\{f_m\}$ is an increasing sequence of functions in $\mathcal{M}(X, \mathcal{X})^+$ (Proposition 1.9) that converges pointwise to the function $f\chi_E$, with $E = \bigcup_n E_n$ in $\mathcal{X}$. Therefore, by the Monotone Convergence Theorem,

$$\mu\left(\bigcup_n E_n\right) = \int_E f \, d\mu = \int f\chi_E \, d\mu = \lim_m \int f_m \, d\mu = \lim_m \int \sum_{n=1}^m f\chi_{E_n} \, d\mu.$$

A trivial induction using the additivity in item (b) ensures that the integral of a finite sum equals the finite sum of each integral, so

$$\lim_m \int \sum_{n=1}^m f\chi_{E_n} \, d\mu = \lim_m \sum_{n=1}^m \int f\chi_{E_n} \, d\mu,$$

and this completes the proof of (c); that is, $\mu\left(\bigcup_n E_n\right) = \sum_n \lambda(E_n)$. Indeed,

$$\lim_m \sum_{n=1}^m \int f\chi_{E_n} \, d\mu = \lim_m \sum_{n=1}^m \int_{E_n} f \, d\mu = \lim_m \sum_{n=1}^m \lambda(E_n) = \sum_n \lambda(E_n). \quad \square$$

Remarks: Carefully note that the Monotone Convergence Theorem deals with functions that are possibly extended real-valued and with measures that are not necessarily finite (not even σ-finiteness is assumed), so infinite integrals and infinite limits are allowed. For instance, if λ is the Lebesgue measure on $\mathfrak{R}$ and $f_n = \chi_{[0,n)}$ for each positive integer n, then $\{f_n\}$ is an increasing sequence of functions in $\mathcal{M}(\mathbb{R}, \mathfrak{R})^+$ with finite integral ($\int f_n \, d\lambda = n$ for each n) that converges pointwise to the function $f = \chi_{[0,\infty)}$ that has an infinite integral. However, the real-valued sequence $\{\int f_n \, d\mu\}$ is unbounded (and so it does not converge in $\mathbb{R}$) but it has the limit $+\infty$ in $\overline{\mathbb{R}}$. Therefore, $\int f \, d\mu = \lim_n \int f_n \, d\mu = +\infty$. We shall see later in this chapter that the Monotone Convergence Theorem still holds if pointwise convergence is weakened to almost everywhere convergence and it survives without monotonicity by assuming just convergence from below (Corollary 3.10). In fact, even convergence from below can be dismissed if we impose uniform convergence and finite measure (Problem 3.6).

3.3 Extension Corollaries

Fatou's Lemma can be viewed as a classic consequence of the Monotone Convergence Theorem, which will help us to prove an extension of the theorem that does not assume monotonicity. Recall that $\underline{f} = \liminf_n f_n$ is a measurable function for any sequence $\{f_n\}$ of extended real-valued measurable functions (Proposition 1.8), and that, if $\{f_n\}$ converges pointwise to a function f (i.e, if $f = \lim_n f_n$), then $f = \underline{f}$.

Lemma 3.6. (Fatou's Lemma). *Let* $(X, \mathcal{X}, \mu)$ *be a measure space. If* $\{f_n\}$ *is a sequence of functions in* $\mathcal{M}(X, \mathcal{X})^+$, *then*

$$\int \underline{f} \, d\mu \leq \liminf_n \int f_n \, d\mu.$$

Proof. Take a sequence $\{f_n\}$ of functions in $\mathcal{M}(X, \mathcal{X})^+$. For each n set

$$\phi_n = \inf_{n \leq k} f_k.$$

That is, each $\phi_n \colon X \to \overline{\mathbb{R}}$ is a function defined by $\phi_n(x) = \inf_{n \leq k} f_k(x)$ for every $x \in X$. Proposition 1.8 ensures that each ϕ_n is measurable, and hence each ϕ_n lies in $\mathcal{M}(X, \mathcal{X})^+$. Observe that, by its very definition, $\{\phi_n\}$ is an increasing sequence. Moreover, it converges pointwise to $\underline{f}$. Indeed,

$$\lim_n \phi_n(x) = \lim_n \inf_{n \leq k} f_k(x) = \liminf_n f_n(x) = \underline{f}(x)$$

for every $x \in X$. Thus the Monotone Convergence Theorem ensures that

$$\int \underline{f} \, d\mu = \lim_n \int \phi_n \, d\mu.$$

Since $\phi_n \leq f_k$ for all $k \geq n$, it follows that $\int \phi_n \, d\mu \leq \int f_k \, d\mu$ for all $k \geq n$, and hence $\int \phi_n \, d\mu \leq \inf_{n \leq k} \int f_k \, d\mu$, so

$$\lim_n \int \phi_n \, d\mu \leq \lim_n \inf_{n \leq k} \int f_k \, d\mu = \liminf_n \int f_n \, d\mu. \qquad \square$$

Proposition 3.7(a) is an application of Fatou's Lemma that gives the first symptom of what will make the basis for defining a new concept of equality in the spaces L^p, through equivalence classes, as we shall see in Chapter 5.

Proposition 3.7. *Let* $(X, \mathcal{X}, \mu)$ *be a measure space. If* f *is a function in* $\mathcal{M}(X, \mathcal{X})^+$ *and* λ *is the measure on* $\mathcal{X}$ *defined in* Proposition 3.5(c), *then*

(a) $f = 0$ μ-*almost everywhere if and only if* $\int f \, d\mu = 0$,

(b) $\lambda(E) = 0$ *for every* $E \in \mathcal{X}$ *such that* $\mu(E) = 0$.

Proof. Let $(X, \mathcal{X}, \mu)$ be a measure space and take any $f \in \mathcal{M}(X, \mathcal{X})^+$.

(a) Since f is a measurable function,

$$E_n = \left\{ x \in X \colon \tfrac{1}{n} < f(x) \right\}$$

is a measurable set such that $\frac{1}{n} \mathcal{X}_{E_n} \leq f$, and hence (Proposition 3.5(a))

$$0 \leq \tfrac{1}{n} \mu(E_n) = \tfrac{1}{n} \int_{E_n} d\mu = \int \tfrac{1}{n} \mathcal{X}_{E_n} \, d\mu \leq \int f \, d\mu$$

for each n. If $\int f\, d\mu = 0$, then $\mu(E_n) = 0$ for all n, and so (since $0 \le f$)

$$\mu(\{x \in X \colon f(x) \ne 0\}) = \mu(\{x \in X \colon 0 < f(x)\}) = \mu\left(\bigcup_n E_n\right) = 0,$$

which means that $f = 0$ μ-almost everywhere. Conversely, set $E = \bigcup_n E_n$ in $\mathcal{X}$ and put $f_n = n\,\chi_E$ for each n. Clearly, $\{f_n\}$ is a sequence of functions in $\mathcal{M}(X, \mathcal{X})^+$ converging pointwise to $f_\infty = +\infty\,\chi_E \colon X \to \overline{\mathbb{R}}$, and hence $\underline{f} = \liminf_n f_n = \lim_n f = f_\infty$. Moreover, $0 \le f \le f_\infty = \underline{f}$. Therefore,

$$0 \le \int f\, d\mu \le \int \underline{f}\, d\mu \le \liminf_n \int f_n\, d\mu$$

according to Fatou's Lemma. If $f = 0$ μ-almost everywhere, then $\mu(E) = 0$ so that $\int f_n\, d\mu = n \int \chi_E\, d\mu = \mu(E) = 0$ for all n. Thus $\liminf_n \int f_n\, d\mu = 0$, and hence $\int f\, d\mu = 0$ by the preceding inequality.

(b) Suppose $\mu(E) = 0$ for some $E \in \mathcal{X}$. Since $\int \chi_E\, d\mu = \mu(E)$, it follows by item (a) that $\chi_E = 0$, and so $f\chi_E = 0$, μ-almost everywhere. Let $\lambda \colon \mathcal{X} \to \overline{\mathbb{R}}$ be the measure of Proposition 3.5(c). Another application of item (a) yields

$$\lambda(E) = \int_E f\, d\mu = \int f\chi_E\, d\mu = 0. \qquad \square$$

Remarks: The property in Proposition 3.7(b) is commonly referred to by saying that "the measure λ is absolutely continuous with respect to the measure μ" (as we shall see in Chapter 7). Thus Propositions 3.5(c) and 3.7(b) say that *if μ is a measure on $\mathcal{X}$ and f is a function in $\mathcal{M}(X, \mathcal{X})^+$, then the set function λ on $\mathcal{X}$, defined for each $E \in \mathcal{X}$ by*

$$\lambda(E) = \int_E f\, d\mu,$$

is a measure that is absolutely continuous with respect to μ. This is another important consequence of the Monotone Convergence Theorem. A central result of Chapter 7 (the Radon–Nikodým Theorem) asserts the converse: *if μ and λ are σ-finite measures and λ is absolutely continuous with respect to μ, then there is an f in $\mathcal{M}(X, \mathcal{X})^+$ satisfying the above identity.*

Corollary 3.8. *Let $(X, \mathcal{X}, \mu)$ be a measure space. If $\{f_n\}$ is an increasing sequence of functions in $\mathcal{M}(X, \mathcal{X})^+$ that converges almost everywhere to a function $f \colon X \to \overline{\mathbb{R}}$ in $\mathcal{M}(X, \mathcal{X})^+$, then*

$$\int f\, d\mu = \lim_n \int f_n\, d\mu.$$

Proof. If $\{f_n\}$ converges μ-almost everywhere to f, then $\{f_n\}$ converges pointwise to f on $E = X \setminus N$ for some $N \in \mathcal{X}$ with $\mu(N) = 0$. Therefore,

$\{f_n \chi_E\}$ is an increasing sequence of functions in $\mathcal{M}(X, \mathcal{X})^+$ converging pointwise to $f \chi_E \in \mathcal{M}(X, \mathcal{X})^+$. By the Monotone Convergence Theorem,

$$\int f \chi_E \, d\mu = \lim_n \int f_n \chi_E \, d\mu.$$

But Propositions 3.5(b) and 3.7(a) ensure (see Problem 3.8) that

$$\int f \chi_E \, d\mu = \int_E f \, d\mu = \int f \, d\mu \quad \text{and} \quad \int f_n \chi_E \, d\mu = \int_E f_n \, d\mu = \int f_n \, d\mu. \quad \square$$

This is the Monotone Convergence Theorem for almost everywhere convergence. It leads to the following version of Fatou's Lemma.

Lemma 3.9. *Let* $(X, \mathcal{X}, \mu)$ *be a measure space. If* $\{f_n\}$ *is a sequence in* $\mathcal{M}(X, \mathcal{X})^+$ *that converges almost everywhere to* $f \in \mathcal{M}(X, \mathcal{X})^+$, *then*

$$\int f \, d\mu \leq \liminf_n \int f_n \, d\mu.$$

Proof. Consider the setup of the previous proof, where $\{f_n \chi_E\}$ converges pointwise to $f \chi_E \in \mathcal{M}(X, \mathcal{X})^+$. Thus Fatou's Lemma says that

$$\int f \chi_E \, d\mu \leq \liminf_n \int f_n \chi_E \, d\mu.$$

But, again, Propositions 3.5(b) and 3.7(a) ensure (see Problem 3.8) that

$$\int f \chi_E \, d\mu = \int f \, d\mu \quad \text{and} \quad \int f_n \chi_E \, d\mu = \int f_n \, d\mu. \quad \square$$

We say that a sequence $\{f_n\}$ *converges from below* to f if $f_n \leq f$ for all n and, of course, it converges to f in some sense. The next extension yields an ultimate version of the Monotone Convergence Theorem that assumes just almost everywhere convergence from below.

Corollary 3.10. *Let* $(X, \mathcal{X}, \mu)$ *be a measure space. If a sequence* $\{f_n\}$ *in* $\mathcal{M}(X, \mathcal{X})^+$ *converges almost everywhere to* $f \in \mathcal{M}(X, \mathcal{X})^+$ *from below, then*

$$\int f \, d\mu = \lim_n \int f_n \, d\mu.$$

Proof. Since $f_n \leq f$, we get $\int f_n \, d\mu \leq \int f \, d\mu$ for all n. Thus, by Lemma 3.9,

$$\int f \, d\mu \leq \liminf_n \int f_n \, d\mu \leq \limsup_n \int f_n \, d\mu \leq \int f \, d\mu. \quad \square$$

3.4   Problems

Problem 3.1. Take a measurable space $(X, \mathcal{X})$ and let "measurable" mean
$\mathcal{X}$-measurable. Show that sum, scalar multiplication, and product of measurable simple functions are measurable simple functions. Thus conclude
that any polynomial $p(\varphi, \psi)$ of measurable simple functions φ and ψ is a
measurable simple function, and $\zeta = \min\{\varphi, \psi\}$ and $\eta = \max\{\varphi, \psi\}$ also
are measurable simple functions. (*Hint:* Proposition 1.5 and Problem 1.3.)

Problem 3.2. First show that the definition of integral of a nonnegative
measurable simple function does not depend on the representation for the
simple function. Now prove Proposition 3.2.

Hint: Homogeneity is readily verified. To prove additivity, verify that

$$\varphi + \psi = \sum_i \sum_j (\alpha_i + \beta_j) \chi_{E_i \cap F_j},$$

where $\varphi = \sum_i \alpha_i \chi_{E_i}$ and $\psi = \sum_j \beta_j \chi_{F_j}$ are canonical representations. So

$$
\begin{aligned}
\int (\varphi + \psi)\, d\mu &= \sum_i \sum_j (\alpha_i + \beta_j) \mu(E_i \cap F_j) \\
&= \sum_i \alpha_i \sum_j \mu(E_i \cap F_j) + \sum_j \beta_j \sum_i \mu(E_i \cap F_j) \\
&= \sum_i \alpha_i \mu(E_i) + \sum_j \beta_j \mu(F_j) = \int \varphi\, d\mu + \int \psi\, d\mu
\end{aligned}
$$

because $\mu(E_i) = \sum_j \mu(E_i \cap F_j)$ and $\mu(F_j) = \sum_i \mu(E_i \cap F_j)$ since $\{E_i\}$ and
$\{F_j\}$ are partitions of X. To prove (c) note that $\varphi \chi_E = \sum_i \alpha_i \chi_{E_i \cap E}$. Thus

$$\lambda(E) = \int \varphi \chi_E\, d\mu = \sum_i \alpha_i \mu(E_i \cap E) = \sum_i \alpha_i \mu_{E_i}(E) \quad \text{for every} \quad E \in \mathcal{X},$$

which is a (finite) linear combination with nonnegative coefficients α_i of
measures μ_{E_i} on $\mathcal{X}$ (cf. Problem 2.11), and hence a measure on $\mathcal{X}$.

Problem 3.3. Let $(X, \mathcal{X}, \mu)$ and $(X, \mathcal{X}, \lambda)$ be measure spaces, E and F
sets in $\mathcal{X}$, and f and g functions in $\mathcal{M}(X, \mathcal{X})^+$. Recall: $\int_E f\, d\mu = \int f \chi_E\, d\mu$;
in particular, $\int f\, d\mu = \int f \chi_X\, d\mu = \int_X f\, d\mu$ (cf. Definition 3.3). Show that

(a) $0 \leq \int_E d\mu = \int \chi_E\, d\mu = \mu(E)$,

(b) $0 \leq \int_E f\, d\mu \leq \int_E g\, d\mu$ whenever $f \leq g$,

(c) $0 \leq \int_E f\, d\mu \leq \int_F f\, d\mu$ whenever $E \subseteq F$,

(d) $0 \leq \int_E f\, d\mu \leq \int_E f\, d\lambda$ whenever $\mu \leq \lambda$.

Problem 3.4. Consider the measurable space $(\mathbb{N}, \wp(\mathbb{N}))$.

(a) Show that every nonnegative function $f: \mathbb{N} \to \overline{\mathbb{R}}$ lies in $\mathcal{M}(\mathbb{N}, \wp(\mathbb{N}))^+$.

Now let μ be the counting measure of Example 2B. Apply Definition 3.3 to verify that, for every nonnegative function $f: \mathbb{N} \to \overline{\mathbb{R}}$,

(b) $\int_E f\, d\mu = \sum_{n \in E} f(n)$ for every $E \in \wp(\mathbb{N})$. Thus $\int f\, d\mu = \sum_{n=1}^{\infty} f(n)$.

Problem 3.5. Take any $x \in \mathbb{R}$, consider the Borel algebra $\mathfrak{R}$ of subsets of $\mathbb{R}$, and let $\delta_x: \mathfrak{R} \to \mathbb{R}$ be the unit point measure concentrated at x of Example 2A. If $f \in \mathcal{M}(\mathbb{R}, \mathfrak{R})^+$, then use Definition 3.3 to show that

(a) $\int_E f\, d\delta_x = f(x)$ if $x \in E$ and is zero otherwise. Thus $\int f\, d\delta_x = f(x)$.

If $\mu = \sum_{n=1}^{m} \alpha_n \delta_n$ with each $\alpha_n \geq 0$, then μ is a measure on $\mathfrak{R}$ for which

(b) $\int f\, d\mu = \sum_{n=1}^{m} \alpha_n \int f\, d\delta_n = \sum_{n=1}^{m} \alpha_n f(n)$.

Problem 3.6. Recall that a sequence $\{f_n\}$ of real-valued functions on X converges uniformly to a real-valued function f on X if, for each $\varepsilon > 0$, there is a positive integer n_ε such that $\sup_{x \in X} |f_n(x) - f(x)| < \varepsilon$ for all $n \geq n_\varepsilon$. Use Problem 3.3 and Proposition 3.5 to prove the following assertion.

○ If $(X, \mathcal{X}, \mu)$ is a *finite* measure space and $\{f_n\}$ is a sequence of real-valued functions in $\mathcal{M}(X, \mathcal{X})^+$ that converges uniformly to a real-valued function f, then f lies in $\mathcal{M}(X, \mathcal{X})^+$ and $\int f\, d\mu = \lim_n \int f_n\, d\mu$.

Hint: Take $k \geq 1$, set $\varepsilon = \frac{1}{k}$, and take $n \geq n_\varepsilon$. Put $E_k = \{x \in X: \frac{1}{k} \leq f(x)\}$ in $\mathcal{X}$. Uniform convergence implies $(f - \frac{1}{k})\chi_{E_k} \leq f_n \leq f + \frac{1}{k}$. Therefore,

$$\int (f - \tfrac{1}{k})\chi_{E_k} d\mu \leq \liminf_n \int f_n\, d\mu \leq \limsup_n \int f_n\, d\mu \leq \int f\, d\mu + \tfrac{1}{k}\mu(X).$$

Now verify that $\int (f - \frac{1}{k})\chi_{E_k} d\mu = \int f\chi_{E_k} d\mu - \frac{1}{k}\mu(E_k)$ because $\mu(X) < \infty$. Since $\int f\chi_{E_k} d\mu = \lambda(E_k)$, where λ is the measure of Proposition 3.5(c), and $\{E_k\}$ is increasing, show that $\lim_k \int f\chi_{E_k} d\mu = \lambda\left(\bigcup_k E_k\right) = \lambda(X) = \int f\, d\mu$ (use Proposition 2.2). Finally, as $\mu(X) < \infty$, check that $\lim_k \frac{1}{k}\mu(E_k) = 0$.

Problem 3.7. Let $(X, \mathcal{X}, \mu)$ be a measure space and let $\{f_n\}$ be a sequence of functions in $\mathcal{M}(X, \mathcal{X})^+$. Use Proposition 3.5(b) to show that

(a) $\int \sum_{n=1}^{m} f_n\, d\mu = \sum_{n=1}^{m} \int f_n\, d\mu$ for each positive integer m.

Now use the Monotone Convergence Theorem (Theorem 3.4) to show that

(b) $\int \sum_{n=1}^{\infty} f_n\, d\mu = \sum_{n=1}^{\infty} \int f_n\, d\mu$.

Problem 3.8. Let $(X, \mathcal{X}, \mu)$ be a measure space, $\{E, F\}$ a measurable partition of X, and $f, g \in \mathcal{M}(X, \mathcal{X})^+$. Use Proposition 3.5(b) to show that

(a) $\int f\,d\mu = \int_E f\,d\mu + \int_F f\,d\mu$.

Take $N \in \mathcal{X}$. Apply Problem 3.3(a) and Proposition 3.7(a) to show that

(b) $\mu(N) = 0$ implies $\int_N f\,d\mu = 0$,

(c) $E = X \setminus N$ and $\mu(N) = 0$ imply $\int_E f\,d\mu = \int f\,d\mu$.

Finally, use Propositions 3.5(b) and 3.7(a) to show that

(d) $\int_E f\,d\mu = \int_E g\,d\mu$ for every $E \in \mathcal{X}$ implies $f = g$ μ-a.e.

> *Hint:* Put $A = \{x \in X : f(x) < g(x)\}$, $B = \{x \in X : f(x) > g(x)\}$, and $C = \{x \in X : f(x) = g(x)\}$. Thus $\{A, B, C\}$ is a measurable partition of X. Since $\int_A f\,d\mu = \int_A g\,d\mu$, and since $(g - f)\chi_A \in \mathcal{M}(X, \mathcal{X})^+$, verify that $\int_A (g - f)\,d\mu = 0$, and so $(g - f) = 0$ μ-a.e. on A. Similarly, $(f - g) = 0$ μ-a.e. on B. Moreover, $f = g$ on C tautologically.

Problem 3.9. Let $\mathcal{X}$ be a σ-algebra of subsets of a set X and let μ be a measure on $\mathcal{X}$. A set $E \in \mathcal{X}$ is σ-*finite* (with respect to μ) if there exists a countable covering of E made up of measurable sets of finite measure (i.e., $E \subseteq \bigcup_n E_n$ with $\mu(E_n) < \infty$ for all n). Now take $f \in \mathcal{M}(X, \mathcal{X})^+$ and use Problem 3.3 and Proposition 3.5 to show that, if $\int f\,d\mu < \infty$, then

(a) $\mu(\{x \in X : f(x) \geq \varepsilon\}) < \infty$ for each $\varepsilon > 0$,

(b) $\mu(\{x \in X : f(x) = +\infty\}) = 0$,

(c) $\{x \in X : f(x) \neq 0\}$ is a σ-finite set.

Hints: (a) If $F_\varepsilon = \{x \in X : \varepsilon \leq f(x)\}$, then $\varepsilon \chi_{F_\varepsilon} \leq f$. (b) Apply Proposition 2.2(d) to $\{F_n\}$. (c) If $E_n = \{x \in X : \frac{1}{n} \leq f(x)\}$, then $\frac{1}{n}\chi_{E_n} \leq f$.

Problem 3.10. On a measure space $(X, \mathcal{X}, \mu)$ let $f \in \mathcal{M}(X, \mathcal{X})^+$ and prove:

○ If $\int f\,d\mu < \infty$, then for every $\varepsilon > 0$ there exists a set $E_\varepsilon \in \mathcal{X}$ such that $\mu(E_\varepsilon) < \infty$ and $\int f\,d\mu \leq \int_{E_\varepsilon} f\,d\mu + \varepsilon$.

Hint: Put $E = \{x \in X : f(x) \neq 0\}$. Verify that $\int f\,d\mu = \int_E f\,d\mu$ (Problem 3.8(a)). Put $E_n = \{x \in X : \frac{1}{n} \leq f(x)\}$. Verify that $\{E_n\}$ is increasing, $E = \bigcup_n E_n$, and $\mu(E_n) = n \int f\,d\mu$ (Problem 3.9(c)). Now apply Theorem 3.4 to show that $\lim_n \int f \chi_{E_n}\,d\mu = \int f\,d\mu$. Thus conclude: for each $\varepsilon > 0$ there is an n_ε such that, with $E_\varepsilon = E_{n_\varepsilon}$, we get $\int f\,d\mu - \int f \chi_{E_\varepsilon}\,d\mu < \varepsilon$.

Problem 3.11. Let $(X, \mathcal{X}, \mu)$ be a measure space. Take f, g in $\mathcal{M}(X, \mathcal{X})^+$ and let λ be the measure on $\mathcal{X}$ such that $\lambda(E) = \int_E f\,d\mu$ for every $E \in \mathcal{X}$ (cf. Proposition 3.5(c)). Show that

$$\int g\,d\lambda = \int gf\,d\mu.$$

This identity is sometimes abbreviated as

$$d\lambda = f\, d\mu,$$

where no independent meaning is assigned to the symbols $d\lambda$ and $d\mu$.

Hint: If $\varphi \in \varPhi_g^+$, that is, if $\varphi = \sum_{i=1}^n \alpha_i \chi_{E_i}$ is a measurable simple function such that $0 \leq \varphi \leq g$, then verify that $\int \varphi\, d\lambda = \int \varphi f\, d\mu = \sum_{i=1}^n \alpha_i \int_{E_i} f\, d\mu$. Now use Problem 1.6 and apply Corollary 3.10.

Problem 3.12. Increasing convergence in Theorem 3.4 was weakened to convergence from below in Corollary 3.10, and this version of the Monotone Convergence Theorem cannot be improved further (even under the assumption of uniform convergence — see Problem 3.6). Indeed, take the *Lebesgue measure space* $(\mathbb{R}, \mathfrak{R}, \lambda)$, where λ is the Lebesgue measure on the Borel algebra $\mathfrak{R}$. Set $f_n = \frac{1}{n}\chi_{[n,\infty)}$ and $g_n = \frac{1}{n}\chi_{[0,n]}$ for each positive integer n and put $f = g = 0$, which are all in $\mathcal{M}(\mathbb{R}, \mathfrak{R})^+$. Show that

(a) $\{f_n\}$ decreases and converges uniformly to f but $\int f_n\, d\lambda = +\infty$ for all n, and so $0 = \int f\, d\lambda \neq \lim_n \int f_n\, d\lambda = +\infty$;

(b) $g \leq g_n$ for all n and $\{g_n\}$, which is not monotone, converges uniformly to g but $\int g_n\, d\lambda = 1$ for all n, and so $0 = \int g\, d\lambda \neq \lim_n \int g_n\, d\lambda = 1$.

Suggested Reading

Bartle [3], Berberian [6], Halmos [13], Royden [22], Rudin [23] (see also [2]).

4
Integrability

4.1 Integrable Functions

Let $(X, \mathcal{X})$ be a measurable space and let $\mathcal{M}(X, \mathcal{X})$ be the collection of all measurable functions on X. A real-valued function f in $\mathcal{M}(X, \mathcal{X})$ can be expressed as $f = f^+ - f^-$, where f^+ and f^- (positive and negative parts of f) are nonnegative measurable functions (Proposition 1.6). We now consider integration of arbitrary functions in $\mathcal{M}(X, \mathcal{X})$ by using the foregoing decomposition. Functions and their integrals are assumed real-valued.

Let $(X, \mathcal{X}, \mu)$ be a measure space and let $\mathcal{L}(X, \mathcal{X}, \mu)$, or simply $\mathcal{L}$ if the measure space is clear in the context, be the collection of all *real-valued functions* from $\mathcal{M}(X, \mathcal{X})$ for which the positive and negative parts have a *finite integral* with respect to the measure μ. That is, $\mathcal{L} = \mathcal{L}(X, \mathcal{X}, \mu) =$

$$\{f : X \to \mathbb{R} : f \text{ is } \mathcal{X}\text{-measurable}, \int f^+ d\mu < \infty \text{ and } \int f^- d\mu < \infty\}.$$

Definition 4.1. Let $(X, \mathcal{X}, \mu)$ be a measure space. The *integral* of a function $f \in \mathcal{L}(X, \mathcal{X}, \mu)$ with respect to μ is the real number

$$\int f \, d\mu = \int f^+ d\mu - \int f^- d\mu.$$

Moreover, the *integral* of f *over* a measurable set E with respect to μ is

$$\int_E f \, d\mu = \int_E f^+ d\mu - \int_E f^- d\mu.$$

An *integrable* (or *μ-integrable*) *function* is precisely a function in $\mathcal{L}(X, \mathcal{X}, \mu)$.

Thus the integral $\int f\,d\mu$ of a *real-valued* function f in $\mathcal{M}(X,\mathcal{X})$ is defined in terms of the integrals $\int f^{\pm}\,d\mu$ of their positive and negative parts $f^{\pm}$ in $\mathcal{M}(X,\mathcal{X})^{+}$, if these integrals (as in Chapter 3) are *finite*. Further notations: $\int f(x)\,d\mu$, $\int f(x)\,d\mu(x)$, or $\int f(x)\,\mu(dx)$. In particular, the *Lebesgue integral* of a measurable function $f\colon\mathbb{R}\to\mathbb{R}$ is the integral defined with respect to Lebesgue measure (λ on $\mathfrak{R}$, or $\bar{\lambda}$ on $\bar{\mathfrak{R}}$; cf. Remark that closes Chapter 2). If the Lebesgue integral of a real-valued function f exists in $\mathbb{R}$, then f is *Lebesgue integrable*. Another notation for the Lebesgue integral: $\int f(x)\,dx$.

Proposition 4.2. *Take an arbitrary measure space* $(X,\mathcal{X},\mu)$.

(a) *If f is a real-valued function in $\mathcal{M}(X,\mathcal{X})$ such that $f=0$ μ-almost everywhere, then f lies in $\mathcal{L}(X,\mathcal{X},\mu)$ and $\int f\,d\mu=0$.*

(b) *If $f\in\mathcal{L}(X,\mathcal{X},\mu)$ and $g\in\mathcal{M}(X,\mathcal{X})$ is bounded, then $fg\in\mathcal{L}(X,\mathcal{X},\mu)$.*

Proof. (a) Take $F_{+}=\{x\in X\colon f(x)>0\}$, $F_{-}=\{x\in X\colon f(x)<0\}$, and $F_{0}=\{x\in X\colon f(x)=0\}$. Now put

$$F^{+}=F_{+}\cup F_{0}=\{x\in X\colon f\geq 0\}\quad\text{and}\quad F^{-}=F_{-}\cup F_{0}=\{x\in X\colon f\leq 0\}.$$

If f lies in $\mathcal{M}(X,\mathcal{X})$, then F^{+} and F^{-} lie in $\mathcal{X}$ (Proposition 1.6) and so F_{+}, F_{-}, and F_{0} also lie in $\mathcal{X}$ (why?). If $f=0$ μ-a.e., then $\mu(F_{+})=0$ (why?). Next recall that $f^{+}=f\chi_{F^{+}}$, and hence $f^{+}=f^{+}\chi_{F^{+}}$. Therefore,

$$\int f^{+}d\mu=\int f^{+}\chi_{F^{+}}\,d\mu=\int_{F^{+}}f^{+}d\mu=\int_{F_{+}}f^{+}d\mu+\int_{F_{0}}f^{+}d\mu=\int_{F_{0}}f^{+}d\mu=0$$

according to Problem 3.8. Similarly, $\int f^{-}d\mu=0$. Thus $f\in\mathcal{L}(X,\mathcal{X},\mu)$ if it is real-valued, and $\int f\,d\mu=0$ by Definition 4.1.

(b) Since $fg=(f^{+}-f^{-})(g^{+}-g^{-})=f^{+}g^{+}+f^{-}g^{-}-f^{+}g^{-}-f^{-}g^{+}$,

$$(fg)^{+}=f^{+}g^{+}+f^{-}g^{-}\quad\text{and}\quad(fg)^{-}=f^{-}g^{+}+f^{+}g^{-}.$$

Since g is bounded, put $\beta=\sup_{x\in X}|g(x)|$ so that $g^{+}\leq\beta$ and $g^{-}\leq\beta$. Thus $f^{+}g^{+}\leq\beta f^{+}$ and hence, by Proposition 3.5(a) and Problem 3.3(b), $\int f^{+}g^{+}d\mu\leq\beta\int f^{+}d\mu<\infty$. Similarly, $\int f^{-}g^{-}d\mu<\infty$, $\int f^{+}g^{-}d\mu<\infty$, and $\int f^{-}g^{+}d\mu<\infty$. Therefore $\int(fg)^{+}d\mu<\infty$ and $\int(fg)^{-}d\mu<\infty$ by Proposition 3.5(b), and so $fg\in\mathcal{L}(X,\mathcal{X},\mu)$. $\qquad\square$

If f_{1} and f_{2} are real-valued in $\mathcal{M}(X,\mathcal{X})^{+}$ with $f_{2}\leq f_{1}$ and $\int f_{2}\,d\mu<\infty$, then $\int(f_{1}-f_{2})\,d\mu=\int f_{1}d\mu-\int f_{2}\,d\mu$. Indeed, write $f_{1}=(f_{1}-f_{2})+f_{2}$ and apply Propositions 1.5 and 3.5(b). We can drop the assumption $f_{2}\leq f_{1}$.

Proposition 4.3. *Let* $(X, \mathcal{X}, \mu)$ *be a measure space. If* f_1 *and* f_2 *are real-valued functions in* $\mathcal{M}(X, \mathcal{X})^+$ *with* $\int f_1 \, d\mu < \infty$ *and* $\int f_2 \, d\mu < \infty$, *then*

$$f_1 - f_2 \in \mathcal{L}(X, \mathcal{X}, \mu) \quad \text{and} \quad \int (f_1 - f_2) \, d\mu = \int f_1 \, d\mu - \int f_2 \, d\mu.$$

Proof. Put $f = f_1 - f_2$ in $\mathcal{M}(X, \mathcal{X})$ (cf. Proposition 1.5). Since f_1, f_2, f^+, and f^- are functions in $\mathcal{M}(X, \mathcal{X})^+$ and $f^+ - f^- = f = f_1 - f_2$, it follows that $f^+ + f_2 = f_1 + f^-$ is in $\mathcal{M}(X, \mathcal{X})^+$, and hence (cf. Proposition 3.5(b))

$$\int f^+ d\mu + \int f_2 \, d\mu = \int f_1 \, d\mu + \int f^- d\mu.$$

Observe that $f^+ \leq f_1$ and $f^- \leq f_2$. Since $\int f_1 \, d\mu < \infty$ and $\int f_2 \, d\mu < \infty$, it follows by Problem 3.3(b) that $\int f^+ d\mu < \infty$ and $\int f^- d\mu < \infty$. Thus $(f_1 - f_2) = f$ lies in $\mathcal{L}(X, \mathcal{X}, \mu)$ and, by Definition 4.1,

$$\int (f_1 - f_2) \, d\mu = \int f \, d\mu = \int f^+ d\mu - \int f^- d\mu = \int f_1 \, d\mu - \int f_2 \, d\mu. \quad \square$$

4.2   Three Fundamental Properties

The first property considered in this section is *absolute integrability*, which says that $|f|$ is integrable if and only if f is. While absolute integrability clearly holds for the Lebesgue integral, it is worth noticing that it does not hold for the Riemann integral.

Indeed, it is a consequence of the well-known result stated in Problem 4.1 that *if a bounded function on a closed and bounded* (i.e., compact) *interval of* $\mathbb{R}$ *has a Riemann integral, then it is* $\overline{\mathfrak{R}}$-*measurable and Lebesgue integrable, and its Riemann and Lebesgue integrals coincide*. However, there exist bounded functions defined on closed and bounded intervals that are Lebesgue but not Riemann integrable. For instance, $f_1(x) = 1$ for $x \in \mathbb{Q}$ and $f_1(x) = -1$ for $x \in \mathbb{R} \backslash \mathbb{Q}$ define a function $f_1 = (2\chi_{\mathbb{Q}} - 1)$ on $[0, 1]$ for which the Riemann integral does not exist but the Lebesgue integral does exist (and is equal to -1). But $|f_1| = 1$, a constant function on $[0, 1]$, is trivially Riemann integrable, and so it is Lebesgue integrable. On the other hand, the situation is different for improper Riemann integrals. If a function either is defined on an unbounded interval or is itself unbounded, then it may have a Riemann integral and not a Lebesgue integral. Sample: $f_2(x) = \frac{\sin x}{x}$ on $[1, \infty)$ defines a Riemann integrable function f_2 (its improper Riemann integral exists and is finite) but $|f_2|$ is not integrable (i.e., it has no finite integral, in any sense) and so f_2 is not Lebesgue integrable.

Lemma 4.4. If $f \in \mathcal{M}(X, \mathcal{X})$, then

(a) $f \in \mathcal{L}(X, \mathcal{X}, \mu)$ if and only if $|f| \in \mathcal{L}(X, \mathcal{X}, \mu)$.

Moreover, if $f \in \mathcal{L}(X, \mathcal{X}, \mu)$, then

(b)
$$\left| \int f \, d\mu \right| \leq \int |f| \, d\mu.$$

Proof. (a) Observe that, for any function $f \colon X \to \mathbb{R}$,

$$|f|^+ = |f| = f^+ + f^- \quad \text{and} \quad |f|^- = 0, \tag{$*$}$$

and recall from Proposition 1.6 that if f is measurable, then so are f^+, f^-, and $|f|$. If f lies in $\mathcal{L}(X, \mathcal{X}, \mu)$, then $\int f^+ d\mu < \infty$ and $\int f^- d\mu < \infty$. Thus we get from $(*)$ and Proposition 3.5(b) that $\int |f|^+ d\mu < \infty$ and $\int |f|^- d\mu < \infty$, and hence $|f|$ lies in $\mathcal{L}(X, \mathcal{X}, \mu)$. Conversely, if $|f|$ lies in $\mathcal{L}(X, \mathcal{X}, \mu)$, then $\int |f|^+ d\mu < \infty$. But $f^+ \leq |f|^+$ and $f^- \leq |f|^+$ by $(*)$. Thus $\int f^+ d\mu < \infty$ and $\int f^- d\mu < \infty$ according to Problem 3.3(b), and so f lies in $\mathcal{L}(X, \mathcal{X}, \mu)$.

(b) If $f \in \mathcal{L}(X, \mathcal{X}, \mu)$, then $f \in \mathcal{M}(X, \mathcal{X},)$ and so $|f| \in \mathcal{L}(X, \mathcal{X}, \mu)$ by (a). Since $|f| = f^+ + f^-$, we get by Proposition 3.5(b) and Definition 4.1 that

$$\left| \int f \, d\mu \right| = \left| \int f^+ d\mu - \int f^- d\mu \right| \leq \int f^+ d\mu + \int f^- d\mu = \int |f| \, d\mu. \quad \square$$

The Monotone Convergence Theorem gave us the first symptoms of linearity for the integral transformation in Proposition 3.5(a,b). Linearity is definitely accomplished in the next lemma.

Lemma 4.5. *$\mathcal{L}$ is a linear space and $\int \colon \mathcal{L} \to \mathbb{R}$ is a linear functional.*

Remarks: What the lemma statement says is twofold. First it says that the collection $\mathcal{L}(X, \mathcal{X}, \mu)$ is a (real) *linear space*. Indeed, since Proposition 1.5 ensures that $\mathcal{M}(X, \mathcal{X})$ is a (real) linear space and since $\mathcal{L}(X, \mathcal{X}, \mu)$ is a subset of $\mathcal{M}(X, \mathcal{X})$, then $\mathcal{L}(X, \mathcal{X}, \mu)$ is a linear space if and only if it is a linear manifold of $\mathcal{M}(X, \mathcal{X})$, which means that *if f and g are functions in $\mathcal{L}(X, \mathcal{X}, \mu)$ and γ is a any real number, then*

(a) γf and $f + g$ lie in $\mathcal{L}(X, \mathcal{X}, \mu)$.

Next consider the real-valued transformation $\int \colon \mathcal{L} \to \mathbb{R}$ that assigns to each function f in $\mathcal{L}(X, \mathcal{X}, \mu)$ the value of its integral $\int f \, d\mu$ in $\mathbb{R}$. This transformation (of the real linear space $\mathcal{L}$ into $\mathbb{R}$) is called a (real) *functional*. The lemma also says that this is a *linear* functional, (i.e., a *homogeneous* and *additive* functional), which means that its domain $\mathcal{L}$ is a linear space and *if f and g are functions in $\mathcal{L}(X, \mathcal{X}, \mu)$ and γ is any real number, then*

(b) $\quad \displaystyle\int \gamma f \, d\mu = \gamma \int f \, d\mu \quad$ and $\quad \displaystyle\int (f + g) \, d\mu = \int f \, d\mu + \int g \, d\mu.$

Thus the proof of Lemma 4.5 is reduced to proving (a) and (b).

Proof. That γf lies in $\mathcal{L}(X, \mathcal{X}, \mu)$ for all $\gamma \in \mathbb{R}$ whenever $f \in \mathcal{L}(X, \mathcal{X}, \mu)$ is a particular case of Proposition 4.2(b). Now, since $-f = f^- - f^+$, we get

$$\gamma f = (\gamma f)^+ - (\gamma f)^- = \begin{cases} |\gamma| f^+ - |\gamma| f^-, & \gamma \geq 0, \\ |\gamma| f^- - |\gamma| f^+, & \gamma < 0, \end{cases}$$

where $|\gamma| f^+$ and $|\gamma| f^-$ lie in $\mathcal{M}(X, \mathcal{X})^+$, with $\int |\gamma| f^+ d\mu = |\gamma| \int f^+ d\mu$ and $\int |\gamma| f^- d\mu = |\gamma| \int f^- d\mu$ (Proposition 3.5(a)), and these integrals are finite because $f \in \mathcal{L}(X, \mathcal{X}, \mu)$. Thus, according to Proposition 4.3,

$$\int \left(\pm |\gamma| f^- - \pm |\gamma| f^- \right) = |\gamma| \left(\pm \int f^+ d\mu - \pm \int f^- d\mu \right),$$

and therefore,

$$\int \gamma f \, d\mu = \begin{cases} |\gamma| \left(\int f^+ d\mu - \int f^- d\mu \right), & \gamma \geq 0, \\ |\gamma| \left(\int f^- d\mu - \int f^+ d\mu \right), & \gamma < 0, \end{cases} \Bigg\} = \gamma \int f \, d\mu.$$

This proves homogeneity. To prove additivity proceed as follows. Take f and g in $\mathcal{L}(X, \mathcal{X}, \mu)$. Since $|f + g| \leq |f| + |g|$, and since $|f + g|$ and $|f| + |g|$ lie in $\mathcal{M}(X, \mathcal{X})^+$ (Propositions 1.5 and 1.6), it follows by Problem 3.3(b) and Proposition 3.5(b) that $\int |f + g| \, d\mu \leq \int (|f| + |g|) \, d\mu = \int |f| \, d\mu + \int |g| \, d\mu$, and hence $|f + g| \in \mathcal{L}(X, \mathcal{X}, \mu)$. Indeed, since f and g are in $\mathcal{L}(X, \mathcal{X}, \mu)$, we get $|f|$ and $|g|$ in $\mathcal{L}(X, \mathcal{X}, \mu)$ by Lemma 4.4, and so $\int |f + g| \, d\mu < \infty$. Another application of Lemma 4.4 ensures that $f + g \in \mathcal{L}(X, \mathcal{X}, \mu)$ (because $f + g \in \mathcal{M}(X, \mathcal{X})$ by Proposition 1.5). Now observe that

$$f + g = (f^+ - f^-) + (g^+ - g^-) = (f^+ + g^+) - (f^- + g^-).$$

Also note that $(f^+ + g^+)$ and $(f^- + g^-)$ lie in $\mathcal{M}(X, \mathcal{X})^+$ by Propositions 1.5 and 1.6 and have finite integrals by Proposition 3.5(b) — because f^+, f^-, g^+, and g^- have finite integrals once f and g lie in $\mathcal{L}(X, \mathcal{X}, \mu)$. Thus Propositions 4.3 and 3.5(b) and Definition 4.1 ensure that

$$\int (f + g) \, d\mu = \int (f^+ + g^+) \, d\mu - \int (f^- + g^-) \, d\mu$$
$$= \int f^+ d\mu - \int f^- d\mu + \int g^+ d\mu - \int g^- d\mu = \int f \, d\mu + \int g \, d\mu. \quad \square$$

The measure generated by nonnegative functions in Proposition 3.5(c) has a natural extension for general functions. This time a signed measure is generated instead, called the *indefinite integral* of f with respect to μ.

Lemma 4.6. *Take an arbitrary integrable function $f \in \mathcal{L}(X, \mathcal{X}, \mu)$. The real-valued set function $\nu \colon \mathcal{X} \to \mathbb{R}$ defined next is a signed measure:*

$$\nu(E) = \int_E f \, d\mu \quad \text{for every} \quad E \in \mathcal{X}.$$

Proof. Since $f = f^+ - f^-$ lies in $\mathcal{L}(X, \mathcal{X}, \mu)$, it follows that the functions f^+ and f^- in $\mathcal{M}(X, \mathcal{X})^+$ have finite integrals. Thus the measures ν^+ and ν^- of Proposition 3.5(c), defined for each $E \in \mathcal{X}$ by

$$\nu^+(E) = \int_E f^+ d\mu \quad \text{and} \quad \nu^-(E) = \int_E f^- d\mu,$$

are finite measures and thus, in particular, signed measures. Recall that

$$\nu(E) = \int_E f \, d\mu = \int_E f^+ \, d\mu - \int_E f^- \, d\mu = \nu^+(E) - \nu^-(E)$$

for each $E \in \mathcal{X}$, and hence $\nu = \nu^+ - \nu^-$. Therefore, as a linear combination of signed measures, ν is itself a signed measure. $\qquad\square$

4.3   The Dominated Convergence Theorem

Here is the most important convergence theorem for integrable functions. It goes in the line of the Monotone Convergence Theorem as in Corollary 3.10, now with no restriction to nonnegative functions, where convergence from below is replaced with dominated convergence.

Theorem 4.7. (Lebesgue Dominated Convergence Theorem). *Take a measure space $(X, \mathcal{X}, \mu)$. Suppose $\{f_n\}$ is a sequence of real-valued functions in $\mathcal{M}(X, \mathcal{X})$ converging almost everywhere to a real-valued function f in $\mathcal{M}(X, \mathcal{X})$, and let g be a nonnegative function in $\mathcal{L}(X, \mathcal{X}, \mu)$. If $|f_n| \leq g$ for all n almost everywhere, then each f_n and f lie in $\mathcal{L}(X, \mathcal{X}, \mu)$ and*

$$\int f \, d\mu = \lim_n \int f_n \, d\mu.$$

Proof. If $|f_n| \leq g$ for all n μ-a.e. and $f_n \to f$ μ-a.e., then $|f| \leq g$ μ-a.e. This implies that each f_n and f are integrable functions according to Problem 4.4(b) — which in fact is a straightforward corollary of Lemma 4.4. Now, since $0 \le g \pm f_n$ and $g \pm f_n \to g \pm f$ μ-a.e., apply Fatou's Lemma (Lemma 3.9) to the sequences $\{g \pm f_n\}$, and use Lemma 4.5 as follows.

$$\int g \, d\mu \pm \int f \, d\mu = \int (g \pm f) \, d\mu \le \liminf_n \int (g \pm f_n) \, d\mu,$$

so

$$\int g \, d\mu + \int f \, d\mu \le \liminf_n \left(\int g \, d\mu + \int f_n \, d\mu \right) = \int g \, d\mu + \liminf_n \int f_n \, d\mu,$$

$$\int g \, d\mu - \int f \, d\mu \le \liminf_n \left(\int g \, d\mu - \int f_n \, d\mu \right) = \int g \, d\mu - \limsup_n \int f_n \, d\mu,$$

and therefore,

$$\limsup_n \int f_n \, d\mu \le \int f \, d\mu \le \liminf_n \int f_n \, d\mu. \qquad \square$$

The Dominated Convergence Theorem plays a key role in the proof of completeness for the space L^p, which is the central result of the next chapter. Applications of it will be frequent throughout the text from now on.

4.4   Problems

Problem 4.1. Consider the following well-known basic result. *A real-valued bounded function defined on a closed and bounded interval of the real line is Riemann integrable if and only if the set of points for which it is not continuous has Lebesgue measure zero.* Let C and S be the Cantor set and the Cantor-like set of Problems 2.9 and 2.10, respectively. Recall that both sets C and S are totally disconnected. Take their characteristic functions $\mathcal{X}_C$ and $\mathcal{X}_S$ from $[0, 1]$ to $\{0, 1\}$. Which of these functions is Riemann integrable? Are they Lebesgue integrable? Which are their integrals?

Problem 4.2. Let $\{f_n\}$ be a sequence of functions in $\mathcal{L}(X, \mathcal{X}, \mu)$. If $\{f_n\}$ converges uniformly to $f \in \mathcal{L}(X, \mathcal{X}, \mu)$ and $\mu(X) < \infty$, then prove that

$$\int f \, d\mu = \lim_n \int f_n \, d\mu,$$

and show that this result may fail without the assumption $\mu(X) < \infty$.

Hint: Verify that if $f - \varepsilon \le f_n \le f + \varepsilon$, then $f^+ - \varepsilon \le f_n^+ \le f^+ + \varepsilon$ and $f^- - \varepsilon \le f_n^- \le f^- + \varepsilon$ for any $\varepsilon > 0$, and conclude that uniform convergence of $\{f_n\}$ to f implies uniform convergence of $\{f_n^+\}$ and $\{f_n^-\}$ to f^+

and f^-, respectively. Now apply Problem 3.6 and Definition 4.1 to prove the identity, and use Problem 3.12(b) to verify that it needs $\mu(X) < \infty$.

Problem 4.3. If a function f lies in $\mathcal{L}(X, \mathcal{X}, \mu)$, then show that its integral is unambiguously defined in terms of the integrals of any pair of functions f_1 and f_2 in $\mathcal{M}(X, \mathcal{X})^+$ with finite integrals such that $f = f_1 - f_2$. Indeed,

$$\int f \, d\mu = \int f^+ \, d\mu - \int f^- \, d\mu = \int f_1 \, d\mu - \int f_2 \, d\mu.$$

Problem 4.4. If f is real-valued in $\mathcal{M}(X, \mathcal{X})$ and $g \in \mathcal{L}(X, \mathcal{X}, \mu)$, then

(a) $f \leq g$ μ-a.e. implies $f \in \mathcal{L}(X, \mathcal{X}, \mu)$ and $\int f \, d\mu \leq \int g \, d\mu$,

(b) $|f| \leq |g|$ μ-a.e. implies $f \in \mathcal{L}(X, \mathcal{X}, \mu)$ and $\int |f| \, d\mu \leq \int |g| \, d\mu$,

(c) $f = g$ μ-a.e. implies $f \in \mathcal{L}(X, \mathcal{X}, \mu)$ and $\int f \, d\mu = \int g \, d\mu$.

Hints: (a) Verify that $f^+ \leq g^+$ and $f^- \leq g^-$ μ-a.e., and apply Problems 3.3(b) and 3.8(c) and Definition 4.1. (b) Use item (a) and Lemma 4.4. (c) By Proposition 4.2 and Lemma 4.5: $\int (f - g) \, d\mu = 0$ and $f = (f - g) + g$.

Problem 4.5. Let f and g be in $\mathcal{L}(X, \mathcal{X}, \mu)$, take any $\gamma \in \mathbb{R}$, let F be an arbitrary measurable set in $\mathcal{X}$, and prove the following assertions.

(a) $\int_F \gamma f \, d\mu = \gamma \int_F f \, d\mu$ and $\int_F (f + g) \, d\mu = \int_F f \, d\mu + \int_F g \, d\mu$.

(b) $\int_E f \, d\mu \geq 0$ $(= 0)$ for all $E \in \mathcal{X}$ if and only if $f \geq 0$ $(= 0)$ μ-a.e.

(c) $\int_E f \, d\mu = \int_E g \, d\mu$ for every $E \in \mathcal{X}$ if and only if $f = g$ μ-a.e.

Hint: $f \chi_E \in \mathcal{L}(X, \mathcal{X}, \mu)$ by Proposition 4.2(b). Use Lemma 4.5 to prove (a) and Propositions 3.7(a) and 4.2(a) and Problem 4.4 to prove (b) and (c).

Problem 4.6. If f and g are in $\mathcal{L}(X, \mathcal{X}, \mu)$, then so are $f \wedge g$ and $f \vee g$. (*Hint:* Problem 1.3, Lemmas 4.4 and 4.5.)

Problem 4.7. A complex-valued function $f : X \to \mathbb{C}$ is measurable if its real and imaginary parts, f_1 and f_2, are real-valued measurable functions (as discussed in Problem 1.7). A measurable complex-valued function $f = f_1 + i f_2$ is *integrable* if its real and imaginary parts are real-valued integrable functions (i.e., if f_1 and f_2 lie in $\mathcal{L}(X, \mathcal{X}, \mu)$). In this case, the *integral* of f is defined as the complex number

$$\int f \, d\mu = \int f_1 \, d\mu + i \int f_2 \, d\mu.$$

Prove the complex version of Lemma 4.4. That is, prove that if f is measurable, then it is integrable if and only if $|f|$ is integrable and, in this a case,

$$\left| \int f \, d\mu \right| \leq \int |f| \, d\mu.$$

Conclude: f is integrable if and only if $|f| \in \mathcal{M}(X, \mathcal{X})$ and $\int |f| \, d\mu < \infty$. Finally, verify the complex version of Problem 4.4(b). That is, if f and g are complex-valued functions, f measurable and g integrable, then $|f| \leq |g|$ μ-a.e. implies that f is integrable and $\int |f| \, d\mu \leq \int |g| \, d\mu$.

Hints: First recall that $|f| \leq |f_1| + |f_2|$, $|f_1| \leq |f|$, and $|f_2| \leq |f|$. Now use Lemmas 4.4 and 4.5 to verify that if f is integrable, then so is $|f_1| + |f_2|$, and hence $|f|$ is integrable by Problem 4.4(b) (since $|f| = (|f_1|^2 + |f_2|^2)^{\frac{1}{2}}$ is measurable — why?). Conversely, if f is measurable and $|f|$ is integrable, use Problem 4.4(b) and Lemma 4.4 to show that f_1 and f_2 are integrable. To prove that $\left| \int f \, d\mu \right| \leq \int |f| \, d\mu$, proceed as follows. Write $\int f \, d\mu = \rho e^{i\theta}$ and put $g = \mathrm{Re}(e^{-i\theta} f)$. Recall that $e^{-i\theta} = \cos\theta - i\sin\theta$ and use Lemma 4.5 to verify that $\left| \int f \, d\mu \right| = e^{-i\theta} \int f \, d\mu = \int e^{-i\theta} f \, d\mu$. Since this integral is nonnegative, conclude that $\left| \int f \, d\mu \right| = \int g \, d\mu = \left| \int g \, d\mu \right|$. Next apply Lemma 4.4 and Problem 3.3(b) to show that $\left| \int g \, d\mu \right| \leq \int |g| \, d\mu \leq \int |f| \, d\mu$.

Problem 4.8. Apply Lemma 4.5 to prove its own complex version. In other words, show that if f and g are complex-valued integrable functions (with respect to a measure μ) and γ is any complex number, then γf and $f + g$ are again integrable complex-valued functions, and

$$\int \gamma f \, d\mu = \gamma \int f \, d\mu \quad \text{and} \quad \int (f + g) \, d\mu = \int f \, d\mu + \int g \, d\mu.$$

Problem 4.9. Prove the following complex version of Theorem 4.7, that is, of the Dominated Convergence Theorem. If $\{f_n\}$ is a sequence of complex-valued measurable functions that converges pointwise to a complex-valued function f, and if g is a nonnegative integrable function (with respect to a measure μ) such that $|f_n| \leq g$ for all n, then f is integrable and

$$\int f \, d\mu = \lim_n \int f_n \, d\mu.$$

Problem 4.10. Verify that the indefinite integral of an integrable function, defined in Lemma 4.6, is countably additive in the following sense. If f lies in $\mathcal{L}(X, \mathcal{X}, \mu)$ and $\{E_n\}$ is a countable partition of any $E \in \mathcal{X}$, then

$$\int_E f \, d\mu = \sum_n \int_{E_n} f \, d\mu.$$

Problem 4.11. Take a measure space $(X, \mathcal{X}, \mu)$. Let $\{f_n\}$ be a sequence of real-valued functions in $\mathcal{M}(X, \mathcal{X})$ such that $\sum_{n=1}^{m} f_n \to f$ almost everywhere to a real-valued function f in $\mathcal{M}(X, \mathcal{X})$, and let g be a nonnegative function in $\mathcal{L}(X, \mathcal{X}, \mu)$. If $\left| \sum_{n=1}^{m} f_n \right| \leq g$ for all m almost everywhere, then f is integrable and

$$\int f \, d\mu = \sum_{n=1}^{\infty} \int f_n \, d\mu.$$

Problem 4.12. Let $\{f_n\}$ be a sequence of functions in $\mathcal{L}(X, \mathcal{X}, \mu)$ and suppose $\sum_{n=1}^{\infty} \int |f_n| \, d\mu < \infty$. Show that the sequence $\left\{ \sum_{n=1}^{m} f_n \right\}$ converges almost everywhere to a function f in $\mathcal{L}(X, \mathcal{X}, \mu)$ and

$$\int f \, d\mu = \sum_{n=1}^{\infty} \int f_n \, d\mu.$$

Hint: The sequence $\left\{ \sum_{n=1}^{m} \int |f_n| \, d\mu \right\}$ of positive numbers is bounded, thus converges in $\mathbb{R}$. Use Problem 3.7 (which is a corollary of the Monotone Convergence Theorem) to conclude that the sequence of positive numbers $\left\{ \int \sum_{n=1}^{m} |f_n| \, d\mu \right\}$ converges in $\mathbb{R}$ and that the sequence $\left\{ \sum_{n=1}^{m} |f_n| \right\}$ of functions in $\mathcal{M}(X, \mathcal{X})^+$ converges pointwise to a function h in $\mathcal{M}(X, \mathcal{X})^+$ such that $\int h \, d\mu = \sum_{n=1}^{\infty} \int |f_n| \, d\mu$. Put $N = \{x \in X : h(x) = +\infty\}$ and set $g(x) = h(x)$ if $x \in X \backslash N$ and $g(x) = 0$ if $x \in N$. Use Problems 3.8 and 3.9 to conclude that the nonnegative (real-valued) function g lies in $\mathcal{L}(X, \mathcal{X}, \mu)$ and the sequence $\left\{ \sum_{n=1}^{m} |f_n| \right\}$ in $\mathcal{M}(X, \mathcal{X})^+$ converges almost everywhere to g, and so the sequence of real-valued functions $\left\{ \sum_{n=1}^{m} f_n \right\}$ in $\mathcal{M}(X, \mathcal{X})$ converges almost everywhere to a function f in $\mathcal{M}(X, \mathcal{X})$ (once it converges absolutely almost everywhere). Verify that $\left| \sum_{n=1}^{m} f_n \right| \leq g$ and apply the Dominated Convergence Theorem to conclude the proof.

Problem 4.13. Take a real-valued function $f \in \mathcal{M}(X, \mathcal{X})$ and, for each positive integer n, consider its n-truncation $f_n \in \mathcal{M}(X, \mathcal{X})$ as defined in Problem 1.5. Show that (a) if $f \in \mathcal{L}(X, \mathcal{X}, \mu)$, then

$$\int f \, d\mu = \lim_{n} \int f_n \, d\mu.$$

Conversely, show that (b) if $\sup_n \int |f_n| \, d\mu < \infty$, then $f \in \mathcal{L}(X, \mathcal{X}, \mu)$.

Hint: Since f is real-valued, verify that $\{f_n\}$ is a sequence of real-valued functions in $\mathcal{M}(X, \mathcal{X})$ that converges pointwise to f and also that $\{|f_n|\}$ is an increasing sequence of real-valued functions in $\mathcal{M}(X, \mathcal{X})^+$ that converges pointwise to $|f|$ (so that $|f_n| \leq |f|$ for all n). To prove (a) show that $|f|$ lies in $\mathcal{L}(X, \mathcal{X}, \mu)$ (why?) and apply the Dominated Convergence Theorem. To prove (b) apply the Monotone Convergence Theorem and conclude that f lies in $\mathcal{L}(X, \mathcal{X}, \mu)$ because $\int |f| \, d\mu < \infty$.

Problem 4.14. Let $\{f_n\}$ be a sequence of functions in $\mathcal{L}(X, \mathcal{X}, \mu)$ and let f be a real-valued function in $\mathcal{M}(X, \mathcal{X})$. Show that

$$\lim_n \int |f_n - f| \, d\mu = 0 \quad \text{implies} \quad f \in \mathcal{L}(X, \mathcal{X}, \mu) \text{ and } \int |f| \, d\mu = \lim_n \int |f_n| \, d\mu.$$

Hint: Since f and f_n are real-valued functions and $\big| |\alpha| - |\beta| \big| \le |\alpha - \beta|$ for all $\alpha, \beta \in \mathbb{R}$, use Problem 3.3(b) and Proposition 3.5(b) to show that if $\lim_n \int |f_n - f| \, d\mu = 0$, then $\limsup_n \int |f_n| \, d\mu \le \int |f| \, d\mu \le \liminf_n \int |f_n| \, d\mu$. Moreover, since $\lim_n \int |f_n - f| \, d\mu = 0$ implies that $|f_n - f|$ lies in $\mathcal{L}(X, \mathcal{X}, \mu)$, and since f_n lies in $\mathcal{L}(X, \mathcal{X}, \mu)$ by hypothesis, apply Lemmas 4.4 and 4.5 to conclude that f lies in $\mathcal{L}(X, \mathcal{X}, \mu)$.

Problem 4.15. Consider all the hypotheses of the Dominated Convergence Theorem. Show that, in addition to the results stated there, we also have

$$\lim_n \int |f_n - f| \, d\mu = 0.$$

Hint: $|f_n - f| \to 0$ and $|f_n - f| \le 2g$ for all n almost everywhere.

Problem 4.16. The dominance assumption (i.e., $|f_n| \le h$ for all n almost everywhere for some $h \in \mathcal{L}(X, \mathcal{X}, \mu)$) cannot be dropped in the Dominated Convergence Theorem (even under the assumption of uniform convergence — see Problem 4.2). Indeed, take the Lebesgue measure space $(\mathbb{R}, \mathfrak{R}, \lambda)$. Put $f = g = 0$ and set $f_n = n \, \chi_{(0, \frac{1}{n}]}$ and $g_n = \frac{1}{n} \chi_{[0,n]}$ for each positive integer n, which are all real-valued functions in $\mathcal{L}(\mathbb{R}, \mathfrak{R}, \lambda)$. Show that

(a) $\{f_n\}$ converges pointwise to f but $0 = \int f \, d\lambda \ne \lim_n \int f_n \, d\lambda = 1$,

(b) $\{g_n\}$ converges uniformly to g but $0 = \int g \, d\lambda \ne \lim_n \int g_n \, d\lambda = 1$.

Suggested Reading

Bartle [3], Berberian [6], Halmos [13], Royden [22], Rudin [23]. For the basic result stated in Problem 4.1, see [21, p. 23] (or [1, p. 206], [27, p. 53]).

5

Spaces L^p

5.1 Normed Spaces

The central idea of this chapter is to provide a topology for the linear space $\mathcal{L}(X, \mathcal{X}, \mu)$ that will actually make it into a Banach space. The first section summarizes just the basics of normed spaces that will be required later. We assume the reader has already been introduced to linear spaces (see Lemma 4.5 and recall that a synonym for *linear space* is *vector space*).

Definition 5.1. Let $\mathbb{F}$ denote either the real field $\mathbb{R}$ or the complex field $\mathbb{C}$, and let $\mathcal{L}$ be an arbitrary linear space over $\mathbb{F}$. A real-valued function

$$\| \ \|\colon \mathcal{L} \to \mathbb{R}$$

is a *norm* on $\mathcal{L}$ if the following conditions are satisfied for all vectors f and g in $\mathcal{L}$ and all scalars γ in $\mathbb{F}$.

(i) $\|f\| \geq 0$ (*nonnegativeness*),

(ii) $\|f\| > 0$ if $f \neq 0$ (*positiveness*),

(iii) $\|\gamma f\| = |\gamma| \|f\|$ (*absolute homogeneity*),

(iv) $\|f + g\| \leq \|f\| + \|g\|$ (*subadditivity — triangle inequality*).

A linear space $\mathcal{L}$ equipped with a norm on it is a *normed space* (synonyms: *normed linear space* and *normed vector space*). If $\mathcal{L}$ is a real or complex linear space (i.e., if $\mathbb{F} = \mathbb{R}$ or $\mathbb{F} = \mathbb{C}$) equipped with a norm on it, then it is referred to as a *real* or *complex normed space,* respectively.

If $\| \ \|$ is a norm on a linear space $\mathcal{L}$, then the resulting normed space is denoted by $(\mathcal{L}, \| \ \|)$ or simply by $\mathcal{L}$ if the norm is clear in the context. Conditions (i)–(iv) are called the *norm axioms*. Note that

$$\|f\| = 0 \quad \text{if and only if} \quad f = 0$$

according to axioms (i) and (ii). If a function $\| \ \| : \mathcal{L} \to \mathbb{R}$ satisfies the three axioms (i), (iii), and (iv) but not necessarily axiom (ii), then it is called a *seminorm* (or a *pseudonorm*); that is, a seminorm surely vanishes at the origin but it may also vanish at a nonzero vector. A linear space $\mathcal{L}$ equipped with a seminorm is called a *seminormed space*.

Definition 5.2. A sequence $\{f_n\}$ of vectors in a normed space $(\mathcal{L}, \| \ \|)$ *converges* to a vector f in $\mathcal{L}$ if for each real number $\varepsilon > 0$ there exists a positive integer n_ε such that

$$n \geq n_\varepsilon \quad \text{implies} \quad \|f_n - f\| < \varepsilon.$$

If $\{f_n\}$ converges to $f \in \mathcal{L}$, then $\{f_n\}$ is said to be a *convergent sequence* and f is said to be the *limit* of $\{f_n\}$. If we need to distinguish it among other convergence modes, then we refer to the preceding concept as *norm convergence* (or *convergence in the norm topology*). A sequence $\{f_n\}$ is a *Cauchy sequence* in $(\mathcal{L}, \| \ \|)$ (or satisfies the *Cauchy criterion*) if for each real number $\varepsilon > 0$ there exists a positive integer n_ε such that

$$n, m \geq n_\varepsilon \quad \text{implies} \quad \|f_m - f_n\| < \varepsilon.$$

A usual notation for the Cauchy criterion is $\lim_{m,n} \|f_m - f_n\| = 0$. We say that $\{f_n\}$ is a *bounded sequence* if $\sup_n \|f_n\| < \infty$.

Proposition 5.3. *Let $(\mathcal{L}, \| \ \|)$ be any normed space.*

(a) *Every convergent sequence in $(\mathcal{L}, \| \ \|)$ is a Cauchy sequence.*

(b) *Every Cauchy sequence in $(\mathcal{L}, \| \ \|)$ is bounded.*

(c) *If a Cauchy sequence in $(\mathcal{L}, \| \ \|)$ has a subsequence that converges in $(\mathcal{L}, \| \ \|)$, then it converges itself in $(\mathcal{L}, \| \ \|)$ and its limit coincides with the limit of that convergent subsequence.*

Proof. Let $(\mathcal{L}, \| \ \|)$ be a normed space and $\{f_n\}$ a sequence of vectors in $\mathcal{L}$.

(a) Take $\varepsilon > 0$. If $\{f_n\}$ converges to $f \in \mathcal{L}$, then there exists $n_\varepsilon \geq 1$ such that $\|f_n - f\| < \frac{\varepsilon}{2}$ for all $n \geq n_\varepsilon$. Since $\|f_m - f_n\| \leq \|f_m - f\| + \|f - f_n\|$ by the triangle inequality, we get $\|f_m - f_n\| < \varepsilon$ whenever $m, n \geq n_\varepsilon$.

(b) If $\{f_n\}$ is Cauchy, then there exists $n_1 > 1$ such that $\|f_m - f_n\| < 1$ for all $m, n \geq n_1$. Let $\beta \in \mathbb{R}$ be the maximum of $\{\|f_m - f_n\| \in \mathbb{R} : m, n < n_1\}$, a finite set, so that $\|f_m - f_n\| \leq \|f_m - f_{n_1}\| + \|f_{n_1} - f_n\| \leq 2 \max\{1, \beta\}$ for all m, n by the triangle inequality. But $\|f_m\| \leq \|f_m - f_1\| + \|f_1\|$ for all m.

(c) Let $\{f_{n_k}\}$ is a subsequence of an Cauchy sequence $\{f_n\}$ that converges to $f \in \mathcal{L}$ (i.e., $\|f_{n_k} - f\| \to 0$ as $k \to \infty$). Take an arbitrary $\varepsilon > 0$. Since $\{f_n\}$ is a Cauchy sequence, there exists a positive integer n_ε such that $\|f_m - f_n\| < \frac{\varepsilon}{2}$ for all $m, n \geq n_\varepsilon$. Since $\{f_{n_k}\}$ converges to f, there exists a positive integer k_ε such that $\|f_{n_k} - f\| < \frac{\varepsilon}{2}$ for all $k \geq k_\varepsilon$. Thus, if j is any integer such that $j \geq k_\varepsilon$ and $n_j \geq n_\varepsilon$ (for instance, if $j = \max\{n_\varepsilon, k_\varepsilon\}$), then $\|f_n - f\| \leq \|f_n - f_{n_j}\| + \|f_{n_j} - f\| < \varepsilon$ for every $n \geq n_\varepsilon$ by the triangle inequality, and therefore $\{f_n\}$ converges to f. $\square$

Although a convergent sequence always is a Cauchy sequence, the converse may fail (as we shall see later). There are, however, many normed spaces with the property that every Cauchy sequence converges. Normed spaces possessing this special property are called *complete*: a normed space $\mathcal{L}$ is complete if every Cauchy sequence in $\mathcal{L}$ is a convergent sequence in $\mathcal{L}$. A *Banach space* is precisely a complete normed space.

Proposition 5.4. *Set* $\mathcal{L} = \mathcal{L}(X, \mathcal{X}, \mu)$. *The function* $\| \ \| : \mathcal{L} \to \mathbb{R}$ *defined by*

$$\|f\| = \int |f| \, d\mu \quad \text{for every} \quad f \in \mathcal{L}(X, \mathcal{X}, \mu)$$

is a seminorm on the linear space $\mathcal{L}(X, \mathcal{X}, \mu)$, *which is such that*

$$\|f\| = 0 \quad \text{if and only if} \quad f = 0 \ \mu\text{-a.e.}$$

Proof. First note that $\mathcal{L}(X, \mathcal{X}, \mu)$ is indeed a real linear space (Lemma 4.5) and that the function $\| \ \| : \mathcal{L} \to \mathbb{R}$ is well defined (i.e., the integral exists in $\mathbb{R}$ for every $f \in \mathcal{L}(X, \mathcal{X}, \mu)$ by Lemma 4.4). Now take arbitrary functions f and g in $\mathcal{L}(X, \mathcal{X}, \mu)$ and any scalar γ in $\mathbb{R}$. Recall that $|\gamma f(x)| = |\gamma||f(x)|$ for every $x \in X$, which means $|\gamma f| = |\gamma||f|$. Also recall that the triangle inequality holds in $\mathbb{R}$, that is, $|\alpha + \beta| \leq |\alpha| + |\beta|$ for any pair $\{\alpha, \beta\}$ of real numbers, and so $|f(x) + g(x)| \leq |f(x)| + |g(x)|$ for every $x \in X$, which means $|f + g| \leq |f| + |g|$. Axiom (i) of Definition 5.1, $\|f\| \geq 0$, is readily verified (Problem 3.3). To verify axioms (iii) and (iv) recall that $\int : \mathcal{L} \to \mathbb{R}$ is a linear functional (Lemma 4.5). Therefore,

$$\|\gamma f\| = \int |\gamma f| \, d\mu = \int |\gamma||f| \, d\mu = |\gamma| \int |f| \, d\mu = |\gamma| \|f\|$$

and (cf. Problem 3.3)

$$\|f+g\| = \int |f+g|\,d\mu \le \int \big(|f|+|g|\big)\,d\mu = \int |f|\,d\mu + \int |g|\,d\mu = \|f\| + \|g\|.$$

Hence $\|\ \|$ is a seminorm on $\mathcal{L}$. Moreover, Proposition 3.7(a) ensures that $\|f\| = 0$ if and only if $f = 0$ μ-a.e. $\square$

However, this seminorm $\|\ \|$ is not a norm on $\mathcal{L}(X, \mathcal{X}, \mu)$. Indeed, it does not satisfy (ii) in Definition 5.1: there may be a function f in $\mathcal{L}(X, \mathcal{X}, \mu)$ such that $f \ne 0$ and $\|f\| = 0$ (e.g., take a Lebesgue integrable function f in $\mathcal{L}(\mathbb{R}, \Re, \lambda)$ such that $f(x) = 0$ for all $x \in \mathbb{R}$ except at the origin, where $f(0) = 1$). In order to make this seminorm into a norm on $\mathcal{L}(X, \mathcal{X}, \mu)$ we need to redefine the concept of equality between functions in $\mathcal{L}(X, \mathcal{X}, \mu)$ (other than the usual pointwise definition) so that axiom (ii) is satisfied.

Let $(X, \mathcal{X}, \mu)$ be a measure space and take arbitrary real-valued functions f and g in $\mathcal{M}(X, \mathcal{X})$. We say that f and g are *equivalent* (or μ-*equivalent*), denoted by $f \sim g$, if $f = g$ almost everywhere (i.e., μ-a.e.). This $\sim$ is indeed an equivalence relation on $\mathcal{M}(X, \mathcal{X})$. Given any *real-valued* function f in $\mathcal{M}(X, \mathcal{X})$, let $[f]$ be the *equivalence class* of f (with respect to μ),

$$[f] = \big\{f' \in \mathcal{M}(X, \mathcal{X}) \colon f' \sim f\big\},$$

which is the subset of $\mathcal{M}(X, \mathcal{X})$ consisting of all functions in $\mathcal{M}(X, \mathcal{X})$ that are μ-equivalent to f. The following necessary and sufficient conditions for equality between equivalence classes are readily verified:

$$[f] = [g] \iff f \sim g \iff f = g \ \mu\text{-a.e.}$$

Recall from Problem 4.4(c) that if f is in $\mathcal{L}(X, \mathcal{X}, \mu)$, then so is every f' in $[f]$ and $\int f'd\mu = \int f\,d\mu$. (Clearly, this implies that if f is in $\mathcal{L}(X, \mathcal{X}, \mu)$, then so is every g' in $[g]$ whenever $[f] = [g]$ and $\int g'd\mu = \int f\,d\mu$.) Now put

$$L^1 = L^1(\mu) = L^1(X, \mathcal{X}, \mu) = \big\{[f] \subseteq \mathcal{M}(X, \mathcal{X}) \colon f \in \mathcal{L}(X, \mathcal{X}, \mu)\big\},$$

the collection of all equivalence classes of function in $\mathcal{L}(X, \mathcal{X}, \mu)$. This collection $L^1(X, \mathcal{X}, \mu)$ is also referred to as the *quotient space* of $\mathcal{L}(X, \mathcal{X}, \mu)$ modulo $\sim$ and also denoted by $\mathcal{L}(X, \mathcal{X}, \mu)/\sim$. Since $\mathcal{L}(X, \mathcal{X}, \mu)$ is a linear space, it can be verified that $L^1(X, \mathcal{X}, \mu)$ is made into a linear space when scalar multiplication and vector addition are defined by

$$\gamma[f] = [\gamma f] \quad \text{and} \quad [f] + [g] = [f+g]$$

for every $[f]$ and $[g]$ in $L^1(X, \mathcal{X}, \mu)$ and every scalar γ. The origin $[0]$ of the linear space $L^1(X, \mathcal{X}, \mu)$ is a subspace,

$$[0] = \big\{f \in \mathcal{L}(X, \mathcal{X}, \mu) \colon f = 0 \ \mu\text{-a.e.}\big\},$$

of the linear space $\mathcal{L}(X,\mathcal{X},\mu)$. Again, $L^1(X,\mathcal{X},\mu)$ is still referred to as the *quotient space* of $\mathcal{L}(X,\mathcal{X},\mu)$ modulo $[0]$ and denoted by $\mathcal{L}(X,\mathcal{X},\mu)/[0]$ as well. The seminorm on $\mathcal{L}(X,\mathcal{X},\mu)$ defined in Proposition 5.4 induces a norm in $L^1(X,\mathcal{X},\mu)$. Indeed, let the function $\|\ \|_1 : L^1 \to \mathbb{R}$ be defined by

$$\|[f]\|_1 = \int |f|\, d\mu \quad \text{for every} \quad [f] \in L^1(X,\mathcal{X},\mu),$$

where $f \in \mathcal{L}(X,\mathcal{X},\mu)$ is any representative of the equivalence class $[f]$.

Proposition 5.5. $\|\ \|_1$ *is a norm on the linear space* $L^1(X,\mathcal{X},\mu)$.

Proof. Let $\|\ \|$ be the seminorm on $\mathcal{L}(X,\mathcal{X},\mu)$ of Proposition 5.4. Note that the function $\|\ \|_1 : L^1 \to \mathbb{R}$ is well defined. Indeed, for any $[f]$ in $L^1(X,\mathcal{X},\mu)$,

$$\|[f]\|_1 = \|f\|,$$

which does not depend on the representative f of the equivalence class $[f]$. Actually, if f and f' are functions in $\mathcal{L}(X,\mathcal{X},\mu)$ such that $f = f'$ μ-a.e., then $|f'| = |f|$ μ-a.e. (since $\big||f| - |f'|\big| \leq |f - f'|$), and so $\int|f'|\,d\mu = \int|f|\,d\mu$ (Problem 4.4(c)). Now observe that $\|\ \|_1$ satisfies all the axioms (i), (ii), (iii), and (iv) of Definition 5.1. In fact, take arbitrary $[f]$ and $[g]$ in $L^1(X,\mathcal{X},\mu)$ and any scalar $\gamma \in \mathbb{R}$ so that, according to Proposition 5.4,

$$\|[f]\|_1 = \|f\| \geq 0,$$

$$\|[f]\|_1 = 0 \iff \|f\| = 0 \iff f = 0 \ \mu\text{-a.e.} \iff [f] = [0],$$

$$\|\gamma[f]\|_1 = \|[\gamma f]\|_1 = \|\gamma f\| = |\gamma|\|f\| = |\gamma|\|[f]\|_1,$$

$$\|[f] + [g]\|_1 = \|[f + g]\|_1 = \|f + g\| \leq \|f\| + \|g\| = \|[f]\|_1 + \|[g]\|_1. \quad \square$$

5.2   The Spaces L^p and L^∞

Let $(X,\mathcal{X},\mu)$ be a measure space. We shall define the linear spaces L^p and L^∞, equip them with norms, and show that they in fact are Banach spaces.

(a)   Take any real number $p \geq 1$. A *real-valued* function f in $\mathcal{M}(X,\mathcal{X})$ is called *p-integrable* if $f^p \in \mathcal{L}(X,\mathcal{X},\mu)$ or, equivalently, if $\int|f|^p d\mu < \infty$ (according to Lemma 4.4). For each $p \geq 1$ put

$$L^p = L^p(\mu) = L^p(X,\mathcal{X},\mu) = \big\{[f] \subseteq \mathcal{M}(X,\mathcal{X}): f^p \in \mathcal{L}(X,\mathcal{X},\mu)\big\},$$

the collection of all equivalence classes of p-integrable functions (i.e., $L^p(X,\mathcal{X},\mu)$ is the collection of all equivalence classes of real-valued functions f in $\mathcal{M}(X,\mathcal{X})$ for which $\int|f|^p d\mu < \infty$ for any representative f of $[f]$), and consider the function $\| \ \|_p \colon L^p \to \mathbb{R}$ defined by

$$\| [f] \|_p = \left(\int |f|^p \, d\mu \right)^{\frac{1}{p}} \quad \text{for every} \quad [f] \in L^p(X, \mathcal{X}, \mu),$$

where f is any representative of the equivalence class $[f]$ in $L^p(X, \mathcal{X}, \mu)$.

(b) We say that a function f in $\mathcal{M}(X, \mathcal{X})$ is *essentially bounded* if it is bounded almost everywhere, that is, if $\sup_{x \in X} |f(x)| < \infty$ μ-a.e., which means that there is a real number $\beta > 0$ such that $|f| \le \beta$ μ-a.e. If a function f in $\mathcal{M}(X, \mathcal{X})$ is essentially bounded, then set

$$\operatorname{ess\,sup} |f| = \inf \{ \beta \ge 0 \colon |f| \le \beta \ \mu\text{-a.e.} \} = \inf_{N \in \mathcal{X}} \ \sup_{x \in X \setminus N} |f(x)|$$

in $\mathbb{R}$, where $\inf_{N \in \mathcal{X}}$ is taken over all $N \in \mathcal{X}$ such that $\mu(N) = 0$. Now put

$$L^\infty = L^\infty(\mu) = L^\infty(X, \mathcal{X}, \mu) = \{ [f] \subseteq \mathcal{M}(X, \mathcal{X}) \colon \sup_{x \in X} |f(x)| < \infty \ \mu\text{-a.e.} \},$$

the collection of all equivalence classes of essentially bounded *real-valued* functions (i.e., $L^\infty(X, \mathcal{X}, \mu)$ is the collection of all equivalence classes of real-valued functions f in $\mathcal{M}(X, \mathcal{X})$ for which $\operatorname{ess\,sup} |f| < \infty$ for any representative f of $[f]$), and consider the function $\| \ \|_\infty \colon L^\infty \to \mathbb{R}$ defined by

$$\| [f] \|_\infty = \operatorname{ess\,sup} |f| \quad \text{for every} \quad [f] \in L^\infty(X, \mathcal{X}, \mu),$$

where f is any representative of the equivalence class $[f]$ in $L^\infty(X, \mathcal{X}, \mu)$.

Since $\mathcal{L}(X, \mathcal{X}, \mu)$ is a linear space, it can be verified that $L^p(X, \mathcal{X}, \mu)$ is made into a linear space if scalar multiplication and vector addition of equivalence classes are defined as in the previous section. Similarly, it is easy to verify that $L^\infty(X, \mathcal{X}, \mu)$ also is made into a linear space under the same definition of scalar multiplication and vector addition.

The elements of $L^p(X, \mathcal{X}, \mu)$ and $L^\infty(X, \mathcal{X}, \mu)$ are equivalence classes of functions (and not functions themselves). However, if f is any function of an equivalence class $[f]$, then it is usual and convienient to write f for $[f]$. We stick to the common usage and refer to a "function f" in $L^p(X, \mathcal{X}, \mu)$ or in $L^\infty(X, \mathcal{X}, \mu)$ instead of "an equivalence class $[f]$ that contains f". Accordingly, we also write $\|f\|_p$ and $\|f\|_\infty$ instead of $\|[f]\|_p$ and $\|[f]\|_\infty$.

Let p and q be real numbers. If $p > 1$ and if $q = \frac{p}{p-1} > 1$ is the unique solution to the equation $\frac{1}{p} + \frac{1}{q} = 1$ (or, equivalently, the unique solution to the equation $p + q = pq$), then p and q are said to be *Hölder conjugates* of each other. Before showing that $\| \ \|_p$ and $\| \ \|_\infty$ actually are norms on the linear spaces L^p and L^∞, respectively, we need the following classic inequalities, which the reader is invited to prove in Problems 5.1 and 5.3.

Proposition 5.6. (Hölder Inequality). *If $f \in L^p$ and $g \in L^q$, where $p > 1$ and $q > 1$ are Hölder conjugates, then $fg \in L^1$ and*

$$\|fg\|_1 \le \|f\|_p \|g\|_q.$$

Moreover, if $f \in L^1$ and $g \in L^\infty$, then $fg \in L^1$ and

$$\|fg\|_1 \le \|f\|_1 \|g\|_\infty.$$

Remark: For $p = q = 2$ we have an important special case, when the Hölder inequality is called the *Schwarz* (or *Cauchy–Schwarz*, or even *Cauchy–Bunyakowski–Schwarz*) *inequality*: if f and g lie in L^2, then $fg \in L^1$ and

$$|\langle f ; g \rangle| \le \int |fg| \, d\mu \le \|f\|_2 \|g\|_2,$$

where $\langle f ; g \rangle = \int fg \, d\mu$ is the *inner product* of f and g in L^2. If $\mu(X) = 1$, then $|\langle f ; \chi_X \rangle| \le \|f\|_1 \le \|f\|_2$, and so $(\int f \, d\mu)^2 \le (\int |f| \, d\mu)^2 \le \int |f|^2 \, d\mu$.

Proposition 5.7. (Minkowski Inequality). *If f and g lie in L^p for some real $p \ge 1$, then $f + g \in L^p$ and*

$$\|f + g\|_p \le \|f\|_p + \|g\|_p.$$

Moreover, if f and g lie in L^∞, then $f + g \in L^\infty$ and

$$\|f + g\|_\infty \le \|f\|_\infty + \|g\|_\infty.$$

Lemma 5.8. $\| \ \|_p$ *and* $\| \ \|_\infty$ *are norms on* L^p *and* L^∞, *respectively.*

Proof. First recall that L^p for each $p \ge 1$ and L^∞ are (real) linear spaces. It was proved that $\| \ \|_1$ is a norm on L^1 in Proposition 5.5. Using exactly the same argument it follows that $\| \ \|_p$ is a norm on L^p for each $p > 1$, where the triangle inequality is precisely the Minkowski inequality of Proposition 5.7. As for $\| \ \|_\infty$ on L^∞, properties (i) and (iii) of Definition 5.1 are trivially verified, (iv) is again the Minkowski inequality of Proposition 5.7, and (ii) follows from the fact that $\operatorname{ess\,sup} |f| = 0$ means $|f| = 0$ μ-a.e. $\square$

5.3   Completeness

The *Completeness Theorem* (upcoming Theorem 5.9) is the statement asserting that the normed spaces L^p and L^∞ (when equipped with the norms $\| \ \|_p$ and $\| \ \|_\infty$) are complete, which means that every Cauchy sequence converges. This central result is also called the *Riesz–Fischer Theorem*.

Theorem 5.9. $(L^p, \| \ \|_p)$ *for each* $p \geq 1$ *and* $(L^\infty, \| \ \|_\infty)$ *are Banach spaces.*

Proof. Consider the normed spaces $(L^p(X, \mathcal{X}, \mu), \| \ \|_p)$ for each real number $p \geq 1$ and $(L^\infty(X, \mathcal{X}, \mu), \| \ \|_\infty)$. We split the proof into two parts.

(a) Take any $p \geq 1$ and consider the normed space $(L^p, \| \ \|_p)$. Let $\{f_n\}$ be an arbitrary Cauchy sequence of functions in L^p so that (cf. Definition 5.2) for any integer $k \geq 1$ there exists another integer $n_k \geq 1$ for which

$$\|f_m - f_n\|_p < \left(\tfrac{1}{2}\right)^k \quad \text{whenever} \quad m, n \geq n_k.$$

This ensures the existence of a subsequence $\{f_{n_k}\}$ of $\{f_n\}$ such that

$$\|f_{n_{k+1}} - f_{n_k}\|_p < \left(\tfrac{1}{2}\right)^k$$

for each $k \geq 1$. Consider the function $g \colon X \to \overline{\mathbb{R}}$ defined, for each $x \in X$, by

$$g(x) = |f_{n_1}(x)| + \sum_{k=1}^{\infty} |f_{n_{k+1}}(x) - f_{n_k}(x)|.$$

Observe that g indeed is a well-defined extended real-valued nonnegative function in $\mathcal{M}(X, \mathcal{X})^+$ (Proposition 1.8). The version of the Monotone Convergence Theorem in Corollary 3.10 ensures that

$$\int g^p \, d\mu \leq \lim_n \int \left(|f_{n_1}| + \sum_{k=1}^{n} |f_{n_{k+1}} - f_{n_k}|\right)^p d\mu.$$

Since each $|f_{n_k}|$ lies in L^p, it follows that $|f_{n_1}| + \sum_{k=1}^{n} |f_{n_{k+1}} - f_{n_k}|$ also lies in the linear space L^p for each integer $n \geq 1$. A trivial induction shows that $\|\sum_{k=1}^{n} g_k\|_p \leq \sum_{k=1}^{n} \|g_k\|_p$ for every positive integer n whenever $\{g_k\}$ is a sequence in L^p by the Minkowski inequality (Proposition 5.7). Therefore,

$$\left(\int \left(|f_{n_1}| + \sum_{k=1}^{n} |f_{n_{k+1}} - f_{n_k}|\right)^p d\mu\right)^{\frac{1}{p}} = \left\||f_{n_1}| + \sum_{k=1}^{n} |f_{n_{k+1}} - f_{n_k}|\right\|_p$$

$$< \|f_{n_1}\|_p + \sum_{k=1}^{n} \left(\tfrac{1}{2}\right)^k = \|f_{n_1}\|_p + 1$$

for all n. By the preceding two inequalities, $\int g^p \, d\mu < (\|f_{n_1}\|_p + 1)^p < \infty$, and so $E = \{x \in X \colon g(x) < \infty\}$ lies in $\mathcal{X}$ with $\mu(X \backslash E) = 0$ according to Problem 3.9(b). This implies that the real series $\sum_{k=1}^{\infty} (f_{n_{k+1}}(x) - f_{n_k}(x))$ converges for every $x \in E$ (since it converges absolutely), and we may set

$$f(x) = \begin{cases} f_{n_1}(x) + \sum_{k=1}^{\infty} \left(f_{n_{k+1}}(x) - f_{n_k}(x)\right) = \lim_k f_{n_{k+1}}(x), & x \in E, \\ 0, & x \notin E, \end{cases}$$

which defines a real-valued function f in $\mathcal{M}(X, \mathcal{X})$ that, in fact, lies in L^p. Indeed, $|f| \leq g$, so

$$\int |f|^p \, d\mu \leq \int g^p \, d\mu < \left(\|f_{n_1}\|_p + 1\right)^p < \infty$$

by Problems 3.3(b), and hence $f \in L^p$. Note that

(i) $\{f_{n_k}\}$ converges almost everywhere to f (i.e., $|f_{n_k} - f| \to 0$ μ-a.e.),

(ii) $|f_{n_k}|^p \leq g^p$ for all k, where g^p is a nonnegative function in $\mathcal{L}(X, \mathcal{X}, \mu)$.

Actually, $f_{n_k}(x) \to f(x)$ for every $x \in E$ and, for all k,

$$|f_{n_k}| = \sum_{j=k}^{\infty} \left(|f_{n_{j+1}}| - |f_{n_j}|\right) \leq \sum_{j=k}^{\infty} |f_{n_{j+1}} - f_{n_j}| \leq \sum_{j=1}^{\infty} |f_{n_{j+1}} - f_{n_j}| \leq g.$$

Moreover, since $|f_{n_k} - f|^p \leq (|f_{n_k}| + |f|)^p \leq (|g| + |f|)^p$ by (ii), and $|f| \leq g$,

(iii) $|f_{n_k} - f|^p \leq (2g)^p$ for all k, where $(2g)^p$ lies in $\mathcal{L}(X, \mathcal{X}, \mu)$.

Thus, since $|f_{n_k} - f|^p \to 0$ μ-a.e. by (i), and according to (iii), it follows by the Dominated Convergence Theorem (Theorem 4.7) that

$$\|f_{n_k} - f\|_p^p = \int |f_{n_k} - f|^p \, d\mu \to 0 \quad \text{as} \quad k \to \infty,$$

so the subsequence $\{f_{n_k}\}$ of $\{f_n\}$ converges in $(L^p, \| \|_p)$ to $f \in L^p$. Apply Proposition 5.3(c) to conclude that the arbitrary Cauchy sequence $\{f_n\}$ converges in $(L^p, \| \|_p)$, and hence the normed space $(L^p, \| \|_p)$ is complete.

(b) Consider the normed space $(L^\infty, \| \|_\infty)$ and let $\{f_n\}$ be any sequence of functions in L^∞. Since a countable collection of sets of measure zero is again a set of measure zero, there exists $N \in \mathcal{X}$ with $\mu(N) = 0$ such that

$$|f_n(x)| \leq \|f_n\|_\infty \quad \text{and} \quad |f_m(x) - f_n(x)| \leq \|f_m - f_n\|_\infty$$

for all $x \in X \backslash N$, for every $m, n \geq 1$. If $\{f_n\}$ is a Cauchy sequence, then for each $\varepsilon > 0$ there exists a positive integer n_ε such that

$$\|f_m - f_n\|_\infty \leq \varepsilon$$

for all $m, n \geq n_\varepsilon$, and therefore

$$\sup_{\substack{x \in X \backslash N \\ m, n \geq n_\varepsilon}} |f_m(x) - f_n(x)| < \varepsilon.$$

This implies that $\{f_n(x)\}$ is a real-valued Cauchy sequence for every x in $X \backslash N$. Since $\mathbb{R}$ (with its usual norm $| \ |$) is a complete normed space, it follows that $\{f_n(x)\}$ converges in $\mathbb{R}$ for every $x \in X \backslash N$. Put

$$f(x) = \begin{cases} \lim_n f_n(x), & x \in X \backslash N, \\ 0, & x \in N, \end{cases}$$

which defines a real-valued function f in $\mathcal{M}(X, \mathcal{X})$ that, in fact, lies in L^∞. Indeed, take x, y in $X \backslash N$ and $n \geq n_\varepsilon$ arbitrary. Observe that

$$|f(x) - f(y)| \leq |f(x) - f_n(x)| + |f_n(x) - f_n(x)| + |f_n(y) - f(y)|.$$

But since the function $|\ |: \mathbb{R} \to \mathbb{R}$ is continuous,

$$|f(x) - f_n(x)| = \left| \lim_m f_m(x) - f_n(x) \right| = \lim_m |f_m(x) - f_n(x)| \leq \varepsilon,$$

and since each f_n lies in L^∞,

$$|f_n(x) - f_n(y)| \leq |f_n(x)| + |f_n(y)| \leq 2\|f_n\|_\infty.$$

Hence we get

$$|f(x)| \leq |f(y)| + |f(x) - f(y)| \leq |f(y)| + 2(\varepsilon + \|f_{n_\varepsilon}\|_\infty),$$

which ensures that $f \in L^\infty$. Moreover, for all $n \geq n_\varepsilon$,

$$\|f - f_n\|_\infty = \sup_{x \in X \backslash N} |f(x) - f_n(x)| \leq \varepsilon.$$

Thus the arbitrary Cauchy sequence $\{f_n\}$ converges in $(L^\infty, \|\ \|_\infty)$ to $f \in L^\infty$, and so the normed space $(L^\infty, \|\ \|_\infty)$ is complete. $\qquad \square$

5.4   Problems

Problem 5.1. Take a measure space $(X, \mathcal{X}, \mu)$ and let f and g be real-valued functions in $\mathcal{M}(X, \mathcal{X})$. If $f^p \in \mathcal{L}(X, \mathcal{X}, \mu)$ and $g^q \in \mathcal{L}(X, \mathcal{X}, \mu)$, where $p > 1$ and $q > 1$ are Hölder conjugates, then $fg \in \mathcal{L}(X, \mathcal{X}, \mu)$ and

$$\int |fg| \, d\mu \leq \left(\int |f|^p \, d\mu \right)^{\frac{1}{p}} \left(\int |g|^q \, d\mu \right)^{\frac{1}{q}}.$$

Hint: Prove the *Young inequality*, namely, $\alpha\beta \leq \frac{\alpha^p}{p} + \frac{\beta^q}{q}$ for every pair of positive real numbers α and β whenever p and q are Hölder conjugates.

If $f \in \mathcal{L}(X, \mathcal{X}, \mu)$ and g is essentially bounded, then $fg \in \mathcal{L}(X, \mathcal{X}, \mu)$ and

$$\int |fg| \, d\mu \leq \text{ess sup} |g| \int |f| \, d\mu.$$

Hint: Proposition 4.2(b) and Problem 4.4 (or Problem 3.8(c)).

Problem 5.2. Generalize the second inequality of Problem 5.1. Take g essentially bounded. If f^p lies in $\mathcal{L}(X, \mathcal{X}, \mu)$ for some $p \geq 1$ (or f is essentially bounded), then $(fg)^p$ lies $\mathcal{L}(X, \mathcal{X}, \mu)$ (fg is essentially bounded) and

$$\int |fg|^p \, d\mu \leq \operatorname{ess\,sup} |g|^p \int |f|^p \, d\mu \quad \left(\operatorname{ess\,sup} |fg| \leq \operatorname{ess\,sup} |f| \operatorname{ess\,sup} |g| \right).$$

Problem 5.3. Suppose $(X, \mathcal{X}, \mu)$ is a measure space. Take any real number $p \geq 1$ and let f and g be p-integrable functions (i.e., real-valued functions in $\mathcal{M}(X, \mathcal{X})$ such that $\int |f|^p \, d\mu < \infty$ and $\int |g|^p \, d\mu < \infty$). Show that $f + g$ is p-integrable (i.e., $\int |f + g|^p \, d\mu < \infty$) and

$$\left(\int |f + g|^p \, d\mu \right)^{\frac{1}{p}} \leq \left(\int |f|^p \, d\mu \right)^{\frac{1}{p}} + \left(\int |g|^p \, d\mu \right)^{\frac{1}{p}}.$$

Hint: This was proved for $p = 1$ in Proposition 5.4. Take any α and β in $\mathbb{R}$. Since $|\alpha + \beta|^p \leq 2^p(|\alpha|^p + |\beta|^p)$, conclude that $\int |f + g|^p \, d\mu < \infty$. Since $|\alpha + \beta|^p \leq (|\alpha| + |\beta|)^p = (|\alpha| + |\beta|)^{p-1}|\alpha| + (|\alpha| + |\beta|)^{p-1}|\beta|$, apply the previous problem (recalling that $(p - 1)q = p$ if q is the Hölder conjugate of p) to prove the displayed inequality.

Also show that if f and g are essentially bounded, then

$$\operatorname{ess\,sup} |f + g| \leq \operatorname{ess\,sup} |f| + \operatorname{ess\,sup} |g|.$$

Problem 5.4. If $f \in L^p(X, \mathcal{X}, \mu)$ for some $p \geq 1$, then for every $\varepsilon > 0$ there exists a measurable simple function φ_ε such that $\|f - \varphi_\varepsilon\|_p < \varepsilon$. This is the so-called *Littlewood second principle*: "every" function is nearly simple.

Hint: Take $0 \leq f \in L^p$. Problem 1.6 ensures the existence of an increasing sequence $\{\varphi_n\}$ of measurable simple functions converging pointwise to f. Thus $(f - \varphi_n)^p \to 0$ pointwise and $0 \leq (f - \varphi_n)^p \leq f^p$. Use the Dominated Convergence Theorem (Theorem 4.7) to conclude that $\|f - \varphi_n\|_p \to 0$.

Problem 5.5. For each $p \geq 1$ let ℓ^p be the set of all scalar-valued (real or complex) sequences $x = \{\xi_k\}$ such that $\sum_{k=1}^{\infty} |\xi_k|^p < \infty$ (i.e., of all scalar-valued p-summable sequences), and let ℓ^∞ be the set of all scalar-valued sequences $x = \{\xi_k\}$ such that $\sup_{k \geq 1} |\xi_k| < \infty$ (i.e., of all scalar-valued *bounded sequences*). It is easy to show that these are (real or complex) linear spaces. Take the measure space $(\mathbb{N}, \wp(\mathbb{N}), \mu)$, where μ is the counting measure of Example 2B. Use Problem 3.4 to show that we may identify

$$\ell^p = L^p(\mathbb{N}, \wp(\mathbb{N}), \mu) \quad \text{and} \quad \ell^\infty = L^\infty(\mathbb{N}, \wp(\mathbb{N}), \mu),$$

where each equivalence class in L^p and L^∞ contains just one element.

Problem 5.6. Apply the previous problem and Lemma 5.8 to verify that

$$\|x\|_p = \Big(\sum_{k=1}^{\infty} |\xi_k|^p\Big)^{\frac{1}{p}} \text{ for every sequence } x = \{\xi_k\} \in \ell^p,$$

$$\|x\|_{\infty} = \sup_{k \geq 1} |\xi_k| \text{ for every sequence } x = \{\xi_k\} \in \ell^{\infty},$$

define norms $\| \ \|_p$ and $\| \ \|_{\infty}$ on ℓ^p and on ℓ^{∞}, respectively, and so $(\ell^p, \| \ \|_p)$ for every $p \geq 1$ and $(\ell^{\infty}, \| \ \|_{\infty})$ are Banach spaces according to Theorem 5.9.

Problem 5.7. It is clear that both the Hölder and the Minkowski inequalities (Propositions 5.6 and 5.7) hold for sequences in ℓ^p and ℓ^{∞} with the norms of the previous problem. Now we consider the *Jensen inequality* for sequences. If p and q are real numbers such that $0 < p < q$, then prove that

(a)
$$\Big(\sum_{k=1}^{\infty} |\xi_k|^q\Big)^{\frac{1}{q}} \leq \Big(\sum_{k=1}^{\infty} |\xi_k|^p\Big)^{\frac{1}{p}}$$

for every scalar-valued sequence $x = \{\xi_k\}$ such that $\sum_{k=1}^{\infty} |\xi_k|^p < \infty$.

Hint: Show that $\sum_{k=1}^{\infty} \alpha_k^r \leq \big(\sum_{k=1}^{\infty} \alpha_k\big)^r$ for each real $r \geq 1$ whenever the sequence of *nonnegative* real numbers $\{\alpha_k\}$ is such that $\sum_{k=1}^{\infty} \alpha_k < \infty$.

With q being any real number greater than p, verify that

(b) $1 < p < q$ implies $\ell^1 \subset \ell^p \subset \ell^q \subset \ell^{\infty}$,

Hint: To show that these are proper inclusions, take $\{\frac{1}{k}\} \in \ell^p \backslash \ell^1$ for $p > 1$.

Problem 5.8. Prove the following generalization of the Hölder inequality. Let p, q, and r be real numbers such that $p > r$, $q > r$, $r \geq 1$, and

$$\frac{1}{p} + \frac{1}{q} = \frac{1}{r}.$$

If $f \in L^p$ and $g \in L^q$, then $fg \in L^r$ and

(a)
$$\|fg\|_r \leq \|f\|_p \|g\|_q.$$

Hint: Verify that $\frac{p}{r}$ and $\frac{q}{r}$ are Hölder conjugates.

Now let $(X, \mathcal{X}, \mu)$ be a *finite* measure space (i.e., $\mu(X) < \infty$) and show that

(b) $1 < r < p$ implies $L^{\infty} \subseteq L^p \subseteq L^r \subseteq L^1$.

Hint: $\|f\|_r \leq \|f\|_p \, \mu(X)^{\frac{p-r}{pr}}$ by the *generalized Hölder inequality* in (a).

Problem 5.9. Let $(X, \mathcal{X}, \mu)$ be a measure space. Show that the following assertions are pairwise equivalent.

(a) $\mu(X) < \infty$.

(b) $L^\infty \subseteq L^p$ for every $p \geq 1$.

(c) $L^\infty \subseteq L^p$ for some $p \geq 1$.

Problem 5.10. If $(X, \mathcal{X}, \mu)$ is a *finite* measure space and $f \in L^\infty$, then

$$\lim_{p \to \infty} \|f\|_p = \|f\|_\infty.$$

Hint: $|f|^p \leq |f|^{p-1}|f|$ implies $\int |f|^p \, d\mu \leq \|f\|_\infty^{p-1} \int |f| \, d\mu$, and hence $\|f\|_p \leq \|f\|_\infty^{1-1/p} \|f\|_1^{1/p} = \|f\|_\infty (\|f\|_1/\|f\|_\infty)^{1/p}$. Then $\limsup \|f\|_p \leq \|f\|_\infty$. For γ in $(0, \|f\|_\infty)$ and $E = \{x \in X \colon |f(x)| > \gamma\}$, $\mu(E)^{1/p} \gamma \leq \left(\int_E |f|^p \, d\mu \right)^{1/p} \leq \|f\|_p$, and so $\gamma \leq \liminf \|f\|_p$. Thus $\|f\|_\infty = \sup \gamma \leq \liminf \|f\|_p$.

Problem 5.11. Take any measure space $(X, \mathcal{X}, \mu)$. Let r and q be such that $1 \leq r \leq q$. If $f \in L^r \cap L^q$, then $f \in L^p$ for every p such that $r \leq p \leq q$.

Hint: Put $E = \{x \in X \colon |f(x)| \leq 1\}$ and $F = \{x \in X \colon |f(x)| > 1\}$ in $\mathcal{X}$ (Proposition 1.6). Show that $\int_E |f|^p \, d\mu + \int_F |f|^p \, d\mu \leq \int_E |f|^r \, d\mu + \int_F |f|^q \, d\mu$.

Problem 5.12. Consider the Lebesgue measure space $(\mathbb{R}, \mathfrak{R}, \lambda)$, let E be any $\mathfrak{R}$-measurable set, consider the restriction $\lambda|_E$ of the Lebesgue measure λ to the σ-algebra $\mathcal{E} = \wp(E) \cap \mathfrak{R}$ of all Borel subsets of E as in Problem 2.11, and take any real number $p \geq 1$. The *Lebesgue spaces* are

$$L^p(E) = L^p(E, \mathcal{E}, \lambda|_E) \quad \text{and} \quad L^\infty(E) = L^\infty(E, \mathcal{E}, \lambda|_E).$$

Put $E = [0, 1]$ and show that the inclusions of Problem 5.8(b) are proper inclusions (i.e., $L^\infty([0, 1]) \subset L^p([0, 1]) \subset L^r([0, 1]) \subset L^1([0, 1])$ if $1 < r < p$) by exhibiting functions in

$$L^2([0, 1]) \backslash L^\infty([0, 1]) \quad \text{and} \quad L^1([0, 1]) \backslash L^2([0, 1]).$$

However, for an unbounded E we get the opposite. Exhibit functions in

$$L^\infty([1, \infty)) \backslash L^2([1, \infty)) \quad \text{and} \quad L^2([1, \infty)) \backslash L^1([1, \infty)).$$

Hints: $\int \frac{dx}{\sqrt{x}} = 2\sqrt{x}$, $\int \frac{dx}{x} = \log(x)$, and $\int \frac{dx}{x^2} = -\frac{1}{x}$.

Problem 5.13. Let $C[0, 1]$ denote the set of all real-valued *continuous* functions $f \colon [0, 1] \to \mathbb{R}$ defined on the closed and bounded interval $[0, 1]$.

Verify that $C[0, 1]$ is a linear space when vector addition and scalar multiplication are pointwise defined. Show that the functions $\| \ \|_p : C[0, 1] \to \mathbb{R}$ for each real number $p \geq 1$ and $\| \ \|_\infty : C[0, 1] \to \mathbb{R}$, given by

$$\|f\|_p = \int_{[0,1]} f(x)\, dx \quad \text{and} \quad \|f\|_\infty = \max_{x \in [0,1]} |f(x)|,$$

are well defined for every $f \in C[0, 1]$ and, in fact, are norms on $C[0, 1]$ (cf. Minkowski inequality). It makes no difference whether the preceding integral is Riemann or Lebesgue. (Why?) Now prove the following assertions.

(a) The normed space $(C[0, 1], \| \ \|_p)$ is not complete for any $p \geq 1$.

   Hint: Consider the $C[0, 1]$-valued sequence $\{f_n\}$ given by

$$f_n(x) = \begin{cases} 1, & x \in [0, \frac{1}{2}], \\ n + 1 - 2nt, & x \in [\frac{1}{2}, \frac{n+1}{2n}], \\ 0, & x \in [\frac{n+1}{2n}, 1]. \end{cases}$$

   Verify the $\{f_n\}$ is a Cauchy sequence in $(C[0, 1], \| \ \|_p)$ but does not converge in $(C[0, 1], \| \ \|_p)$ to any (continuous) function in $C[0, 1]$.

(b) $(C[0, 1], \| \ \|_\infty)$ is a Banach space. (*Hint:* Proof of Theorem 5.9(b).)

Problem 5.14. Take the Lebesgue Space $L^p([0, 1])$ of all p-integrable functions on $[0, 1]$ as defined in Problem 5.12. Let $R^p[0, 1]$ be the subset of $L^p[0, 1]$ consisting of all Riemann integrable functions f for which $|f|^p$ has a finite Riemann integral. It is easy to show that each $R^p[0, 1]$ is a linear manifold of $L^p[0, 1]$, and hence a linear space. Equip each of them with the norm $\| \ \|_p$ and consider the normed spaces $(R^p[0, 1], \| \ \|_p)$ for every $p \geq 1$. Now let $\{f_n\}$ be a sequence of real-valued functions on $[0, 1]$ defined by

$$f_n(x) = \begin{cases} 1, & x = \frac{k}{n!} \in [0, 1] \text{ for some integer } k \geq 0, \\ 0, & \text{otherwise.} \end{cases}$$

Show that each f_n lies in $R^p[0, 1]$ for every $p \geq 1$. (*Hint:* Problem 4.1.) Use the same argument to show that the Dirichlet function f on $[0, 1]$,

$$f(x) = \chi_{[0,1] \cap \mathbb{Q}}(x) = \begin{cases} 1, & x \in [0, 1] \cap \mathbb{Q}, \\ 0, & x \in [0, 1] \backslash \mathbb{Q}, \end{cases}$$

lies in $L^p[0, 1] \backslash R^p[0, 1]$ for any $p \geq 1$. (*Hint:* $[0, 1] \backslash \mathbb{Q}$ is totally disconnected and of measure 1 — cf. Problems 2.7(b) and 4.1.) Finally, verify that

$$f_n \to f \quad \text{pointwise.}$$

Hint: If $f_n(x) = 1$, then $x = \frac{k}{n!}$, so $x = \frac{k(n+1)}{(n+1)!}$ and hence $f_{n+1}(x) = 1$.

Can you infer from this problem that $(R^p[0,1], \| \ \|_p)$ is not complete?

Problem 5.15. As in the previous problem, let $(R^p[0,1], \| \ \|_p)$ be the linear manifold of the Banach space $(L^p[0,1], \| \ \|_p)$. The purpose of this problem is to show that the normed space $(R^p[0,1], \| \ \|_p)$ is not complete for any $p \geq 1$. Consider the decreasing collection $\{S_n\}$ of closed subsets of $S_0 = [0,1]$ used to build up the Cantor-like set $S = \bigcap_n S_n$ of Problem 2.10. Note that

$$\lambda(S_n) = 1 - \sum_{i=0}^{n-1} \frac{2^i}{4^{i+1}} = \frac{1}{2} + \frac{1}{2^{n+1}}$$

is the length of S_n for each $n \geq 1$. Now consider the sequence $\{f_n\}$ of characteristic functions of S_n for every $n \geq 1$,

$$f_n(x) = \chi_{S_n}(x) = \begin{cases} 1, & x \in S_n, \\ 0, & x \in S_0 \backslash S_n. \end{cases}$$

Note that each f_n belongs to $R^p[0,1]$ for every $p \geq 1$. (Why? See Problem 4.1.) Finally, let f be the characteristic function of S; that is,

$$f(x) = \chi_S(x) = \begin{cases} 1, & x \in S, \\ 0, & x \in S_0 \backslash S. \end{cases}$$

(a) Show that $\{f_n\}$ is a Cauchy sequence in $(R^p[0,1], \| \ \|_p)$.

Hint: Verify that $\|f_m - f_n\|_p^p \leq \frac{1}{2^{m+1}}$ whenever $m \leq n$.

(b) Show that $f \in L^p[0,1]$ and $\{f_n\}$ converges in $(L^p[0,1], \| \ \|_p)$ to f.

Hint: Verify that $\|f_n - f\|_p^p = \lambda(S_n \backslash S) = \frac{1}{2^{n+1}}$ since $f_n - f = \chi_{S_n \backslash S}$.

(c) Show that $f \notin R^p[0,1]$. (*Hint:* Problems 2.10 and 4.1.)

Use (a), (b), and (c) to conclude that, for any $p \geq 1$, the normed space $(R^p[0,1], \| \ \|_p)$ is not a Banach space; that is,

$$(R^p[0,1], \| \ \|_p) \text{ is an incomplete normed space.}$$

Suggested Reading

Bartle [3], Halmos [13], Royden [22], Rudin [23]. See also [6], [7], [10], [11], [27], [28] and, for an introduction to Banach spaces, e.g., [17, Chapter 4].

6
Convergence

6.1 Previous Concepts

This chapter deals with sequences $\{f_n\}$ of *real-valued* functions $f_n\colon X \to \mathbb{R}$ on X. We have met four convergence concepts so far. These are reviewed in this section. Further concepts will be considered in subsequent sections.

Definition 6.1. A sequence of functions $\{f_n\}$ *converges pointwise* to a real-valued function $f\colon X \to \mathbb{R}$ if the real-valued sequence $\{f_n(x)\}$ converges to $f(x)$ in $\mathbb{R}$ for every $x \in X$. That is, if $|f_n(x) - f(x)| \to 0$ as $n \to \infty$ for every $x \in X$. In other words, $\{f_n\}$ converges pointwise to f if for every $\varepsilon > 0$ and every $x \in X$ there is a positive integer $n_{\varepsilon,x}$ such that

$$n \geq n_{\varepsilon,x} \quad \text{implies} \quad |f_n(x) - f(x)| < \varepsilon.$$

In such a case we write $f_n \to f$ pointwise. If this $n_{\varepsilon,x}$ does not depend on x, then $\{f_n\}$ *converges uniformly* to f, and we write $f_n \to f$ uniformly. That is, a sequence of functions $\{f_n\}$ converges uniformly to a real-valued function $f\colon X \to \mathbb{R}$ if $\sup_{x \in X} |f_n(x) - f(x)| \to 0$ as $n \to \infty$ or, equivalently, if for every $\varepsilon > 0$ there is a positive integer n_ε such that

$$n \geq n_\varepsilon \quad \text{implies} \quad \sup_{x \in X} |f_n(x) - f(x)| < \varepsilon.$$

Remarks: Pointwise convergence is, in fact, convergence in the normed space $(\mathbb{R}, |\ |)$. Since $(\mathbb{R}, |\ |)$ is a Banach space, a real-valued sequence $\{f_n(x)\}$ converges to $f(x)$ in $\mathbb{R}$ for every $x \in X$ if and only if $\{f_n(x)\}$

is a Cauchy sequence in $(\mathbb{R}, | \ |)$ for every $x \in X$, which means that for every $\varepsilon > 0$ and every $x \in X$ there is a positive integer $n_{\varepsilon,x}$ such that

$$m, n \geq n_{\varepsilon,x} \quad \text{implies} \quad |f_m(x) - f_n(x)| < \varepsilon.$$

Notation: $\lim_{m,n} |f_m(x) - f_n(x)| = 0$ for every $x \in X$. Again, if this $n_{\varepsilon,x}$ does not depend on x, then $\{f_n\}$ is a *uniform Cauchy sequence*, which means that for every $\varepsilon > 0$ there is a positive integer n_ε such that

$$m, n \geq n_\varepsilon \quad \text{implies} \quad \sup_{x \in X} |f_m(x) - f_n(x)| < \varepsilon.$$

Notation: $\lim_{m,n} \sup_{x \in X} |f_m(x) - f_n(x)| = 0$. Clearly, if $f_n \to f$ uniformly, then $\{f_n\}$ is a uniform Cauchy sequence. In fact, for each m and n,

$$\sup_{x \in X} |f_m(x) - f_n(x)| \leq \sup_{x \in X} |f_m(x) - f(x)| + \sup_{x \in X} |f(x) - f_n(x)|.$$

Conversely, if $\{f_n\}$ is a uniform Cauchy sequence, then $\{f_n(x)\}$ is a real-valued Cauchy sequence, and so it converges to, say, $f(x)$ for every $x \in X$. This defines a function $f \colon X \to \mathbb{R}$ such that $f_n \to f$ uniformly. Indeed, take $x \in X$ and $\varepsilon > 0$ arbitrary so that there is a positive integer n_ε for which

$$m \geq n_\varepsilon \quad \text{implies} \quad |f_m(x) - f(x)| = \lim_n |f_m(x) - f_n(x)| < \varepsilon.$$

Example 6A. $f_n \to f$ uniformly $\overset{\Rightarrow}{\not\Leftarrow}$ $f_n \to f$ pointwise.

It is plain that uniform convergence implies pointwise convergence (to the *same* and unique limit), and it is readily verified that the converse fails. Actually, for each integer $n \geq 1$ let $f_n \colon [0,1] \to \mathbb{R}$ be given by $f_n(x) = x^n$ if $x \in [0,1)$ and $f_n(x) = 0$ if $x = 1$. Thus the sequence $\{f_n\}$ converges pointwise to the null function 0 (i.e., $0(x) = 0$ for all $x \in [0,1]$) but it does not converge to 0 uniformly (since $\sup_{x \in [0,1]} |f_n(x)| = 1$ for all $n \geq 1$), and hence it does not converge uniformly to any function.

The foregoing convergence concepts are measure free. The next two require a measure space. Thus from now on we fix a measure space $(X, \mathcal{X}, \mu)$, and by a "measurable function" we mean an $\mathcal{X}$-measurable function on X.

Definition 6.2. A sequence functions $\{f_n\}$ converges *almost everywhere* with respect to the measure $\mu \colon \mathcal{X} \to \overline{\mathbb{R}}$ (or μ-*almost everywhere*) to a real-valued function $f \colon X \to \mathbb{R}$ if the real-valued sequence $\{f_n(x)\}$ converges to $f(x)$ in $\mathbb{R}$ for every x except in a set of measure zero. In this case we write $f_n \to f$ μ-a.e. (or simply a.e. if the measure μ is clear in the context or immaterial). Formally, $\{f_n\}$ converges almost everywhere to f if there exists a set $N \in \mathcal{X}$ with $\mu(N) = 0$ such that for every $\varepsilon > 0$ and every $x \in X \backslash N$ there is a positive integer $n_{\varepsilon,x}$ for which

$$n \geq n_{\varepsilon,x} \quad \text{implies} \quad |f_n(x) - f(x)| < \varepsilon.$$

A sequence $\{f_n\}$ *converges in* $L^p(X, \mathcal{X}, \mu)$ for some $p \geq 1$ if it converges in the Banach space $(L^p, \|\ \|_p)$. Notation: $f_n \to f$ in L^p. In other words, $\{f_n\}$ converges in L^p if $\|f_n - f\|_p \to 0$ as $n \to \infty$ for some real-valued function $f \in L^p$. That is, for each $\varepsilon > 0$ there is a positive integer n_ε such that

$$n \geq n_\varepsilon \quad \text{implies} \quad \|f_n - f\|_p < \varepsilon.$$

Example 6B. $\{f_n\}$ converges pointwise $\overset{\Longrightarrow}{\underset{\not\Longleftarrow}{}}$ $\{f_n\}$ converges a.e.

The first thing to be pointed out is that, for any measure space $(X, \mathcal{X}, \mu)$,

$$f_n \to f \text{ pointwise} \quad \Longrightarrow \quad f_n \to f \ \mu\text{-a.e.},$$

since $\mu(\varnothing) = 0$ for every measure μ on any σ-algebra $\mathcal{X}$. We shall now discuss *how* the converse fails. First recall that the limit of a sequence that converges pointwise or uniformly is unique, and so is the limit of a sequence that converges in L^p, where, in this case, uniqueness is understood almost everywhere (the class of equivalence $[f]$ containing f is unique). However, opposite to the concepts of pointwise convergence, uniform convergence, and convergence in L^p, the concept of almost everywhere convergence, as defined earlier, does not imply uniqueness of the almost everywhere limit. In fact, consider the function $f : \mathbb{R} \to \mathbb{R}$ defined by $f(x) = 0$ for all $x \in \mathbb{R} \backslash \{0\}$ and $f(0) = 1$. Let $\{f_n\}$ be a constant sequence such that $f_n = f$ for all $n \geq 1$. Take the Lebesgue measure space $(\mathbb{R}, \mathfrak{R}, \lambda)$. Trivially, $\{f_n\}$ converges pointwise to f (which is its unique pointwise limit), and $\{f_n\}$ converges a.e. to any function $f' : \mathbb{R} \to \mathbb{R}$ such that $f' = f$ a.e. (i.e., $f'(x) = f(x)$ for every $x \in \mathbb{R} \backslash N$ for some $N \in \mathfrak{R}$ such that $\lambda(N) = 0$). In particular, it converges almost everywhere to the null function $0 : \mathbb{R} \to \mathbb{R}$ (since $\lambda(\{0\}) = 0$ and $f(x) = 0$ for all $x \in \mathbb{R} \backslash \{0\}$), and also to f itself (since $\lambda(\varnothing) = 0$). Thus, $f_n \to 0$ λ-a.e. and $f_n \to f \neq 0$ pointwise. The reader might argue that such an example would be meaningless had we decided to work with equivalence classes of functions,

$$[f] = \{f' : \mathbb{R} \to \mathbb{R} : f' = f \ \lambda\text{-a.e.}\},$$

rather than with functions themselves. This would in fact yield uniqueness for the almost everywhere limit but would not be enough to avoid the failure of the converse; it fails anyway. Actually, if $g_n = 1^{-n} f$, then $g_n \to 0$ λ-a.e. and $\{g_n\}$ does not converge pointwise (to any function).

In general, uniform convergence and convergence in L^p are not related,

$$\{f_n\} \text{ converges uniformly} \quad \not\!\!\!\Longrightarrow\!\!\!\!\Longleftarrow \quad \{f_n\} \text{ converges in } L^p$$

(see Problem 6.1). However, if $\{f_n\}$ converges both uniformly and in L^p, then the limits coincide almost everywhere, and in a *finite* measure space uniform convergence implies convergence in L^p to the same (a.e.) limit.

Proposition 6.3. *Let $\{f_n\}$ be a sequence of functions in L^p for some $p \geq 1$ and let f, f', and f'' be real-valued measurable functions.*

(a) *If $f_n \to f'$ uniformly and $f_n \to f''$ in L^p, then $f' = f''$ μ-a.e.*

(b) *If $\mu(X) < \infty$ and $f_n \to f$ uniformly, then $f \in L^p$ and $f_n \to f$ in L^p.*

Proof. Let f_n, f, f', and f'' be real-valued $\mathcal{X}$-measurable functions on X.

(a) $|f' - f''| \leq |f' - f_n| + |f_n - f''| \leq \sup_{x \in X} |f'(x) - f_n(x)| + |f_n - f''|$ for each n. Since $f_n \to f'$ uniformly it follows that $|f' - f''| \leq |f_n - f''|$ for all n. Since $f_n \to f''$ in L^p, it also follows (cf. Problem 3.3(b)) that

$$0 \leq \int |f' - f''|^p \, d\mu \leq \int |f_n - f''|^p \, d\mu = \|f_n - f''\|_p^p \to 0,$$

and therefore $f' = f''$ μ-a.e. (cf. Propositions 1.5, 1.6, and 3.7(a)).

(b) Take an arbitrary real $p \geq 1$. If each f_n lies in L^p, then

$$\|f_m - f_n\|_p^p = \int |f_m - f_n|^p \, d\mu \leq \sup_{x \in X} |f_m(x) - f_n(x)|^p \, \mu(X)$$

for any positive integers m and n. Since $\{f_n\}$ is uniformly Cauchy (because $f_n \to f$ uniformly) and since $\mu(X) < \infty$, it follows that $\{f_n\}$ is a Cauchy sequence in the Banach space $(L^p, \|\,\|_p)$, and so it converges in L^p; and the L^p limit coincides a.e. with the uniform limit f (i.e., it is $[f]$) by (a). $\square$

Note that, according to Problem 6.2(a), even if $\mu(X) < \infty$,

$$\{f_n\} \text{ converges pointwise} \quad \not\!\!\!\Longrightarrow \quad \{f_n\} \text{ converges in } L^p.$$

However, under the assumption of *dominated convergence*, just convergence almost everywhere is enough to ensure convergence in L^p.

Proposition 6.4. *Let $\{f_n\}$ be a sequence of functions in L^p for some $p \geq 1$, let f be real-valued measurable function, and take g in L^p.*

$$|f_n| \leq g \text{ for all } n \text{ and } f_n \to f \text{ } \mu\text{-a.e.} \implies f \in L^p \text{ and } f_n \to f \text{ in } L^p.$$

Proof. First note that the dominance assumption, viz. $|f_n| \leq g$ for all n, is equivalent to almost everywhere dominance, namely $|f_n| \leq g$ for all n μ-a.e., once the functions f_n and g are in L^p, where inequalities (as well as equalities) are understood in the sense of equivalence classes, and so they are in fact always interpreted almost everywhere. Since $|f_n|^p \leq g^p$ for all n μ-a.e. and $f_n \to f$ μ-a.e., it follows that $|f|^p \leq g^p$ μ-a.e. Thus, if $g \in L^p$, then f^p is integrable (cf. Problem 4.4(b)); that is, $f \in L^p$. Moreover,

$$|f_n - f|^p \leq \left(|f_n| + |f|\right)^p \leq 2g^p \in L^1 \quad \text{and} \quad |f_n - f|^p \to 0 \ \mu\text{-a.e.}$$

Then $f_n \to f$ in L^p by the Dominated Convergence Theorem (Theorem 4.7):

$$\lim_n \|f_n - f\|_p^p = \lim_n \int |f_n - f|^p \, d\mu = 0. \qquad \square$$

Remark: Every constant function lies in any L^p if the measure is *finite*. This yields the following *uniformly bounded* version of the previous proposition.

$$\sup_n |f_n| < \infty, \ \mu(X) < \infty \ \text{and} \ f_n \to f \ \mu\text{-a.e.} \implies f_n \to f \in L^p \ \text{in} \ L^p.$$

Convergence in L^p is of paramount importance since it is convergence in the norm topology of the Banach space $(L^p, \| \ \|_p)$. In general, it does not imply uniform convergence, nor it is implied by uniform convergence (Problem 6.1), but in a finite measure space it is weaker than uniform convergence (Proposition 6.3). Even in a finite measure space it is not implied by almost everywhere convergence (not even by pointwise convergence — Problem 6.2(a)), but under the dominance hypothesis it becomes weaker than almost everywhere convergence (Proposition 6.4). On the other hand,

$$\{f_n\} \text{ converges in } L^p \ \not\Longrightarrow \ \{f_n\} \text{ converges a.e.,}$$

even under finite measure and dominance condition (Problem 6.2(b)). However, convergence in measure (next section) is weaker than L^p-convergence.

6.2 Convergence in Measure

Again, take a fixed measure space $(X, \mathcal{X}, \mu)$ and let $\{f_n\}$ be a sequence of *real-valued measurable* functions (i.e., f_n lies in $\mathcal{M}(X, \mathcal{X})$ for each n).

Definition 6.5. A sequence $\{f_n\}$ of real-valued functions in $\mathcal{M}(X, \mathcal{X})$ *converges in measure* to a real-valued function f in $\mathcal{M}(X, \mathcal{X})$ if

$$\lim_n \mu\left(\{x \in X : |f_n(x) - f(x)| \geq \alpha\}\right) = 0$$

for every $\alpha > 0$. The sequence $\{f_n\}$ is *Cauchy in measure* if, for every $\alpha > 0$,

$$\lim_{m,n} \mu\big(\{x \in X \colon |f_m(x) - f_n(x)| \geq \alpha\}\big) = 0.$$

Remarks: Observe that, for every integer $n \geq 1$ and each real $\alpha > 0$, the set

$$F_n(\alpha) = \big\{x \in X \colon |f_n(x) - f(x)| \geq \alpha\big\}$$

lies in $\mathcal{X}$. (Why?) Thus convergence in measure (notation: $f_n \to f$ in measure) means $\lim_n \mu(F_n(\alpha)) = 0$. This implies that the sequence $\{\mu(F_n(\alpha))\}$ is eventually real-valued. Hence $f_n \to f$ in measure if and only if for each $\varepsilon > 0$ and each $\alpha > 0$ there is a positive integer $n_{\varepsilon,\alpha}$ such that

$$n \geq n_{\varepsilon,\alpha} \quad \text{implies} \quad \mu\big(\{x \in X \colon |f_n(x) - f(x)| \geq \alpha\}\big) < \varepsilon.$$

Similarly, the sequence $\{f_n\}$ is Cauchy in measure if and only if for each $\varepsilon > 0$ and each $\alpha > 0$ there exists a positive integer $n_{\varepsilon,\alpha}$ such that

$$m, n \geq n_{\varepsilon,\alpha} \quad \text{implies} \quad \mu\big(\{x \in X \colon |f_m(x) - f_n(x)| \geq \alpha\}\big) < \varepsilon.$$

Example 6C. $\quad f_n \to f$ in $L^p \quad \overset{\Longrightarrow}{\nLeftarrow} \quad f_n \to f$ in measure.

Convergence in L^p implies convergence in measure to the same limit since

$$\alpha^p \mu\big(F_n(\alpha)\big) = \int_{F_n(\alpha)} \alpha^p \, d\mu \leq \int_{F_n(\alpha)} |f_n - f|^p \, d\mu \leq \int |f_n - f|^p \, d\mu = \|f_n - f\|_p$$

for each $\alpha > 0$ and every $n \geq 1$. The converse fails even if $\mu(X) < \infty$. Indeed, it is readily verified that the sequence $\{f_n\}$ of Problem 6.2(a), which does not converge in L^p, converges in measure to the null function 0.

Example 6D. $\quad f_n \to f$ uniformly $\quad \overset{\Longrightarrow}{\nLeftarrow} \quad f_n \to f$ in measure.

Actually, if $f_n \to f$ uniformly, then for any $\varepsilon > 0$ there is an n_ε such that

$$n \geq n_\varepsilon \implies \sup_{x \in X} |f_n(x) - f(x)| < \varepsilon \implies F_n(\varepsilon) = \varnothing \implies \mu\big(F_n(\varepsilon)\big) = 0,$$

which means that the sequence $\{\mu(F_n(\alpha))\}$ not only converges to zero but is in fact eventually null for each $\alpha > 0$. However, the converse fails. For instance, the sequence $\{f_n\}$ of Problem 6.1(b) converges in L^p, and hence it converges in measure but does not converge uniformly. It is worth noticing that $\{f_n\}$, as in Problem 6.1(a), also yields another example of a sequence that converges in measure (because it converges uniformly) but not in L^p.

Proposition 6.6. *Take a sequence $\{f_n\}$ of real-valued measurable functions.*

(a) *If $\{f_n\}$ converges in measure, then it is Cauchy in measure.*

(b) *If $\{f_n\}$ converges in measure, then the limit is unique μ-a.e.*

(c) *If $\{f_n\}$ is Cauchy in measure and has a subsequence that converges in measure, then it converges in measure itself and its limit coincides with the limit of that subsequence.*

Proof. Let f_n, f, and f' be real-valued measurable functions. Put $F_n(\alpha) = \{x \in X : |f_n(x) - f(x)| \geq \alpha\}$ and $F'_n(\alpha) = \{x \in X : |f_n(x) - f'(x)| \geq \alpha\}$.

(a) Since $|f_m(x) - f_n(x)| \leq |f_m(x) - f(x)| + |f_n(x) - f(x)|$, it follows that $\{x \in X : |f_m(x) - f_n(x)| \geq \alpha\} \subseteq F_m(\frac{\alpha}{2}) \cup F_n(\frac{\alpha}{2})$, and hence

$$\mu\big(\{x \in X : |f_m(x) - f_n(x)| \geq \alpha\}\big) \leq \mu\big(F_m(\tfrac{\alpha}{2})\big) + \mu\big(F_n(\tfrac{\alpha}{2})\big).$$

If $\{f_n\}$ converges in measure to f, then $\lim_n \mu(F_n(\frac{\alpha}{2})) = 0$, and therefore $\lim_{m,n} \mu(\{x \in X : |f_m(x) - f_n(x)| \geq \alpha\}) = 0$, for every $\alpha > 0$.

(b) Since $|f(x) - f'(x)| \leq |f_n(x) - f(x)| + |f_n(x) - f'(x)|$, it follows that $\{x \in X : |f(x) - f'(x)| \geq \alpha\} \subseteq F_n(\frac{\alpha}{2}) \cup F'_n(\frac{\alpha}{2})$, and hence

$$\mu\big(\{x \in X : |f(x) - f'(x)| \geq \alpha\}\big) \leq \mu\big(F_n(\tfrac{\alpha}{2})\big) + \mu\big(F'_n(\tfrac{\alpha}{2})\big).$$

If $\{f_n\}$ converges in measure to both f and f', then $\lim_n \mu(F_n(\frac{\alpha}{2})) = \lim_n \mu(F'_n(\frac{\alpha}{2})) = 0$, so $\mu(\{x \in X : |f(x) - f'(x)| \geq \alpha\}) = 0$ for every $\alpha > 0$, and therefore $f' = f$ μ-a.e. (which follows by the inclusion $\{x \in X : |f(x) - f'(x)| > 0\} \subseteq \bigcup_{k=1}^{\infty}\{x \in X : |f(x) - f'(x)| \geq \frac{1}{k}\}$).

(c) Let $\{f_{n_k}\}$ be a subsequence of a sequence $\{f_n\}$. Take $\alpha > 0$ and $\varepsilon > 0$ arbitrary. If $\{f_n\}$ is Cauchy in measure, then there exists an integer $n_{\varepsilon,\alpha}$ such that $\mu\{x \in X : |f_m(x) - f_n(x)| \geq \frac{\alpha}{2}\} < \frac{\varepsilon}{2}$ for $m, n \geq n_{\varepsilon,\alpha}$. If $\{f_{n_k}\}$ converges in measure to f, then there exists an integer $k_{\varepsilon,\alpha}$ such that $\mu\{x \in X : |f_{n_k}(x) - f(x)| \geq \frac{\alpha}{2}\} < \frac{\varepsilon}{2}$ for $k \geq k_{\varepsilon,\alpha}$. Thus if j is an integer such that $j \geq k_{\varepsilon,\alpha}$ and $n_j \geq n_{\varepsilon,\alpha}$, then

$$\mu\big\{x \in X : |f_n(x) - f_{n_j}(x)| \geq \tfrac{\alpha}{2}\big\} < \tfrac{\varepsilon}{2}$$

and

$$\mu\big\{x \in X : |f_{n_j}(x) - f(x)| \geq \tfrac{\alpha}{2}\big\} < \tfrac{\varepsilon}{2}$$

for $n \geq n_{\varepsilon,\alpha}$. Since $|f_n(x) - f(x)| \leq |f_n(x) - f_{n_j}(x)| + |f_{n_j}(x) - f(x)|$,

$$\{x \in X : |f_n(x) - f(x)| \geq \alpha\}$$
$$\subseteq \big\{x \in X : |f_n(x) - f_{n_j}(x)| \geq \tfrac{\alpha}{2}\big\} \cup \big\{x \in X : |f_{n_j}(x) - f(x)| \geq \tfrac{\alpha}{2}\big\},$$

and hence

$$\mu(\{x \in X : |f_n(x) - f(x)| \geq \alpha\})$$
$$\leq \mu(\{x \in X : |f_n(x) - f_{n_j}(x)| \geq \tfrac{\alpha}{2}\}) + \mu(\{x \in X : |f_{n_j}(x) - f(x)| \geq \tfrac{\alpha}{2}\})$$

so that $\mu(\{x \in X : |f_n(x) - f(x)| \geq \alpha\}) < \varepsilon$ for all $n \geq n_{\varepsilon,\alpha}$. Thus $\lim_n \mu(F_n(\alpha)) = 0$, and therefore $\{f_n\}$ converges in measure to f. $\square$

Convergences in measure and almost everywhere are not related. In fact,

$$\{f_n\} \text{ converges in measure} \quad \not\Rightarrow \quad \{f_n\} \text{ converges a.e.,}$$

even if $\mu(X) < \infty$. For instance, the sequence $\{f_n\}$ of Problem 6.2(b) acts on a finite measure space, converges in measure (since it converges in L^p) to the null function 0, but $\{f_n(x)\}$ fails to converge for any x, and hence $\{f_n\}$ does not converge a.e. (and so it does not converge pointwise). Although uniform convergence implies convergence in measure (see Example 6D),

$$\{f_n\} \text{ converges pointwise} \quad \not\Rightarrow \quad \{f_n\} \text{ converges in measure}$$

(in particular, convergence a.e. does not imply convergence in measure). Indeed, the sequence $\{f_n\}$ of Problem 6.3(a,b) converges pointwise (i.e., everywhere in $\mathbb{R}$, and hence a.e.) to the null function 0, but it is not Cauchy in measure and so it does not converge in measure (Proposition 6.6(a)). Recall that not converging in measure implies that it does not converge both in L^p and uniformly (Examples 6C and 6D). However, convergence almost everywhere implies convergence in measure if $\mu(X) < \infty$, as we shall see in Propositions 6.12 and 6.13.

Proposition 6.7. *If a sequence of real-valued measurable functions is Cauchy in measure, then it has a subsequence that converges both in measure and almost everywhere.*

Proof. Consider a measure space $(X, \mathcal{X}, \mu)$ and let $\{f_n\}$ be a sequence of real-valued functions in $\mathcal{M}(X, \mathcal{X})$. Take any integer $k \geq 1$. If $\{f_n\}$ is Cauchy in measure, then there exists another integer $n_k \geq 1$ for which

$$\mu(\{x \in X : |f_m(x) - f_n(x)| \geq (\tfrac{1}{2})^k\}) < (\tfrac{1}{2})^k \quad \text{whenever} \quad m, n \geq n_k.$$

This ensures the existence of a subsequence $\{f_{n_k}\}$ of $\{f_n\}$ such that

$$\mu(\{x \in X : |f_{n_{k+1}}(x) - f_{n_k}(x)| \geq (\tfrac{1}{2})^k\}) < (\tfrac{1}{2})^k.$$

For each $k \geq 1$, set

$$E_k = \bigcup_{j=k}^{\infty} \{x \in X : |f_{n_{j+1}}(x) - f_{n_j}(x)| \geq (\tfrac{1}{2})^j\}$$

in $\mathcal{X}$ so that $\mu(E_k) < \sum_{j=k}^{\infty} \left(\frac{1}{2}\right)^j = \left(\frac{1}{2}\right)^{k-1}$. Put $N = \bigcap_{k=1}^{\infty} E_k$, a set in $\mathcal{X}$ with $\mu(N) = 0$ (reason: $N \subseteq E_k$ and so $\mu(N) \le \mu(E_k)$, for all $k \ge 1$). Since

$$f_{n_i}(x) - f_{n_j}(x) = \sum_{\ell=1}^{i-j} f_{n_{i-\ell+1}}(x) - f_{n_{i-\ell}}(x) = \sum_{\ell=j}^{i-1} f_{n_{\ell+1}}(x) - f_{n_\ell}(x)$$

for every $1 \le j < i$, we get: if $x \in X \backslash E_k$ and $k \le j < i$, then

$$|f_{n_i}(x) - f_{n_j}(x)| \le \sum_{\ell=j}^{i-1} |f_{n_{\ell+1}}(x) - f_{n_\ell}(x)| < \sum_{\ell=j}^{i-1} \left(\tfrac{1}{2}\right)^\ell < \left(\tfrac{1}{2}\right)^{j-1}.$$

Therefore, if $x \in X \backslash N = X \backslash \bigcap_{k=1}^{\infty} E_k = \bigcup_{k=1}^{\infty} (X \backslash E_k)$ or, equivalently, if $x \in X \backslash E_k$ for some $k \ge 1$, then the above inequality holds for every pair of distinct integers $i, j \ge k$, which has two consequences.

(i) First, it implies that $\{f_{n_j}(x)\}$ is a Cauchy sequence in $\mathbb{R}$, and hence it converges in $\mathbb{R}$, for every $x \in X \backslash N$. Since $\mu(N) = 0$, it follows that

$$f_{n_j} \to f \text{ a.e.}, \quad \text{where} \quad f(x) = \begin{cases} \lim_j f_{n_j}(x), & x \in X \backslash N, \\ 0, & x \in N, \end{cases}$$

which defines a the real-valued measurable function f on X.

(ii) Moreover, take any $k \ge 1$. Since $f_{n_j}(x) \to f(x)$ for every $x \in X \backslash E_k$, the foregoing inequality also implies that, for every $j \ge k$,

$$|f_{n_j}(x) - f(x)| = \lim_i |f_{n_i}(x) - f_{n_j}(x)| \le \left(\tfrac{1}{2}\right)^{j-1} \le \left(\tfrac{1}{2}\right)^{k-1}.$$

Now take arbitrary $\alpha > 0$ and $\varepsilon > 0$. Since $\mu(E_k) < \left(\frac{1}{2}\right)^{k-1}$ for each $k \ge 1$, it follows that there exists an integer $k' = k_{\varepsilon,\alpha} \ge 1$ for which

$$\mu(E_{k'}) < \left(\tfrac{1}{2}\right)^{k'-1} < \min\{\varepsilon, \alpha\},$$

and hence, according to the inequality in (ii), for every $j \ge k'$,

$$\{x \in X : |f_{n_j}(x) - f(x)| \ge \alpha\} \subseteq \{x \in X : |f_{n_j}(x) - f(x)| \ge \left(\tfrac{1}{2}\right)^{k'}\} \subseteq E_k'.$$

Thus

$$\mu(\{x \in X : |f_{n_j}(x) - f(x)| \ge \alpha\}) \le \mu(E_k') < \varepsilon$$

for all $j \ge k' = k_{\varepsilon,\alpha}$, which means that $f_{n_j} \to f$ in measure. $\qquad\square$

The next result is an important consequence of Proposition 6.7, which is referred to as the *Riesz–Weyl Theorem*. It ensures that *a sequence converges in measure if and only if it is Cauchy in measure*.

Theorem 6.8. *If a sequence of real-valued measurable functions is Cauchy in measure, then it converges in measure.*

Proof. Propositions 6.6(c) and 6.7.   □

Recall that convergence in L^p implies convergence in measure, but the converse fails even in a finite measure space (Example 6C). However, another application of Proposition 6.7 ensures that dominated convergence in measure implies convergence in L^p.

Proposition 6.9. *Let $\{f_n\}$ be a sequence of functions in L^p for some $p \geq 1$, let f be real-valued measurable function, and take g in L^p.*

$|f_n| \leq g$ for all n and $f_n \to f$ in measure $\implies f \in L^p$ and $f_n \to f$ in L^p.

Proof. Recall that $|f_n| \leq g$ for all n if and only if $|f_n| \leq g$ for all n μ-a.e. (i.e., in this context, dominance is equivalent to almost everywhere dominance — cf. proof of Proposition 6.4). If $\{f_n\}$ does not converge in L^p to f, then there exist a subsequence $\{g_k\}$ of $\{f_n\}$ and a real number $\varepsilon > 0$ such that

$$\|g_k - f\|_p \geq \varepsilon \quad \text{for every} \quad k \geq 1. \tag{$*$}$$

If $\{f_n\}$ converges in measure to f, then every subsequence of it converges in measure to f. In particular, $\{g_k\}$ converges in measure to f. Propositions 6.6(a) and 6.7 ensure that $\{g_k\}$ has a subsequence $\{g_{k_j}\}$ that converges both in measure and almost everywhere. Since $\{g_{k_j}\}$ converges in measure, Proposition 6.6 ensures that it must converge to f. Since $\{g_{k_j}\}$ also converges almost everywhere to f, and since $|g_{k_j}| \leq g \in L^p$ for all j, it follows by Proposition 6.4 that f lies in L^p and $\{g_{k_j}\}$ converges in L^p to f, which contradicts the assertion in $(*)$. Thus $\{f_n\}$ converges in L^p to f.   □

6.3   Almost Uniform Convergence

Take a fixed measure space $(X, \mathcal{X}, \mu)$, let f be a real-valued function on X, and let $\{f_n\}$ be a sequence of real-valued functions on X. We say that $\{f_n\}$ *converges uniformly almost everywhere* to f if $\{f_n\}$ converges uniformly to f on $X \backslash N$ (i.e., $\lim_n \sup_{x \in X \backslash N} |f_n(x) - f(x)| = 0$) for some set N in $\mathcal{X}$ with $\mu(N) = 0$. Equivalently, if there is a set N in $\mathcal{X}$ with $\mu(N) = 0$ such that for every $\varepsilon > 0$ there exists a positive integer n_ε for which

$$n \geq n_\varepsilon \quad \text{implies} \quad \sup_{x \in X \backslash N} |f_n(x) - f(x)| < \varepsilon.$$

In other words, if the sequence converges uniformly on the complement of a set of measure zero. However, a weaker convergence concept, requiring uniform convergence on the complement of sets that have arbitrarily small measure, will be enough for our purposes. In fact, we have already met this kind of convergence in part (ii) of the proof of Proposition 6.7.

Definition 6.10. A sequence $\{f_n\}$ *converges almost uniformly* to f if for every $\delta > 0$ there is a set E_δ in $\mathcal{X}$ with $\mu(E_\delta) < \delta$ such that $\{f_n\}$ converges uniformly to f on $X\backslash E_\delta$ (i.e., $\lim_n \sup_{x\in X\backslash E_\delta} |f_n(x) - f(x)| = 0$) or, equivalently, if for each $\delta > 0$ and each $\varepsilon > 0$ there exists a set E_δ in $\mathcal{X}$ with $\mu(E_\delta) < \delta$ and a positive integer $n_{\varepsilon,\delta}$ such that

$$n \geq n_{\varepsilon,\delta} \quad \text{implies} \quad \sup_{x\in X\backslash E_\delta} |f_n(x) - f(x)| < \varepsilon.$$

Notation: $f_n \to f$ a.u. A sequence $\{f_n\}$ is *almost uniformly Cauchy* if for every $\delta > 0$ there is a set E_δ in $\mathcal{X}$ with $\mu(E_\delta) < \delta$ such that $\{f_n\}$ is a uniform Cauchy sequence on $X\backslash E_\delta$ (i.e., $\lim_{m,n} \sup_{x\in X\backslash E_\delta} |f_m(x) - f_n(x)| = 0$) or, equivalently, if for each $\delta > 0$ and each $\varepsilon > 0$ there exist a set E_δ in $\mathcal{X}$ with $\mu(E_\delta) < \delta$ and a positive integer $n_{\varepsilon,\delta}$ such that

$$m, n \geq n_{\varepsilon,\delta} \quad \text{implies} \quad \sup_{x\in X\backslash E_\delta} |f_m(x) - f_n(x)| < \varepsilon.$$

Example 6E. Obviously, uniform convergence implies uniform almost everywhere convergence (put $N = \varnothing$), which in turn implies almost uniform convergence (put $E_\delta = N$), and the converses fail even if $\mu(X) < \infty$ — i.e.,

$$f_n \to f \text{ uniformly} \quad \overset{\Longrightarrow}{\not\Leftarrow} \quad f_n \to f \text{ uniformly a.e.} \quad \overset{\Longrightarrow}{\not\Leftarrow} \quad f_n \to f \text{ a.u.}$$

Indeed, consider the *finite* Lebesgue measure space $([0,1], \wp([0,1]) \cap \mathfrak{R}, \lambda)$ and let $\{f_n\}$ be a sequence of real-valued functions on $[0,1]$ such that $f_n(x) = 0$ for all $x \neq 0$ and $f_n(0) = 1^{-n}$ for every $n \geq 1$. Clearly, $\{f_n\}$ does not converge pointwise, and so it does not converge uniformly, but it converges uniformly almost everywhere. It is also plain that the sequence $\{f_n\}$ of Example 6A converges almost uniformly to the null function 0, but it does not converge to 0 uniformly almost everywhere since $\sup_{[0,1]\backslash N} |f_n(x)| = 1$ for any Borel set $N \subseteq [0,1]$ with $\lambda(N) = 0$, for all n.

It is readily verified that if $f_n \to f$ almost uniformly, then $\{f_n\}$ is an almost uniform Cauchy sequence. In fact, for each m and n,

$$\sup_{x\in X\backslash E_\delta} |f_m(x) - f_n(x)| \leq \sup_{x\in X\backslash E_\delta} |f_m(x) - f(x)| + \sup_{x\in X\backslash E_\delta} |f(x) - f_n(x)|.$$

The next result ensures the converse so that *a sequence converges almost uniformly if and only if it is an almost uniform Cauchy sequence.*

Proposition 6.11. *If a sequence $\{f_n\}$ is almost uniformly Cauchy, then it converges almost uniformly and also almost everywhere to the same limit f. Moreover, if each f_n is measurable, then so is the limit function f.*

Proof. Consider a measure space $(X, \mathcal{X}, \mu)$ and let $\{f_n\}$ be an almost uniform Cauchy sequence. Thus for each integer $k \geq 1$ there is a set E_k in $\mathcal{X}$ with $\mu(E_k) < \frac{1}{k}$ such that

$$\lim_{m,n} \sup_{x \in X \setminus E_k} |f_m(x) - f_n(x)| = 0.$$

Put $N = \bigcap_{k=1}^{\infty} E_k$ in $\mathcal{X}$ so that $\mu(N) \leq \mu(E_k) < \frac{1}{k}$ for every $k \geq 1$, and so $\mu(N) = 0$. If $x \in X \setminus N = X \setminus \bigcap_{k=1}^{\infty} E_k = \bigcup_{k=1}^{\infty}(X \setminus E_k)$, then the real-valued sequence $\{f_n(x)\}$ is Cauchy in $\mathbb{R}$, and hence it converges in $\mathbb{R}$. Therefore,

$$f_n \to f \quad \text{a.e.,} \quad \text{where} \quad f(x) = \begin{cases} \lim_n f_n(x), & x \in X \setminus N, \\ 0, & x \in N. \end{cases}$$

This defines a the real-valued function

$$f = \lim_n \left(f_n \chi_{X \setminus N} \right)$$

on X, which is measurable if each f_n is (cf. Propositions 1.8 and 1.9). Moreover, since $\{f_n\}$ is almost uniformly Cauchy and $f_n(x) \to f(x)$ for every x on each $X \setminus E_k \subseteq X \setminus N$, it follows that $f_n(x) \to f(x)$ uniformly on $X \setminus E_k$. Indeed, for $k \geq 1$ and $\varepsilon > 0$ arbitrary there exists $n_{\varepsilon,k} \geq 1$ for which

$$m, n \geq n_{\varepsilon,k} \quad \text{implies} \quad \sup_{x \in X \setminus E_k} |f_m(x) - f_n(x)| < \varepsilon,$$

and hence, for every $x \in X \setminus E_k$,

$$m \geq n_{\varepsilon,k} \quad \text{implies} \quad |f_m(x) - f(x)| = \lim_n |f_m(x) - f_n(x)| < \varepsilon,$$

which in turn implies $\sup_{x \in X \setminus E_k} |f_m(x) - f(x)| < \varepsilon$. This ensures that

$$f_n \to f \quad \text{a.u.}$$

Actually, for each $\delta > 0$, take k large enough so that $\frac{1}{k} \leq \delta$ and put $E_\delta = E_k$ in $\mathcal{X}$ so that $\mu(E_\delta) < \delta$ and $\{f_n\}$ converges uniformly on $X \setminus E_\delta$. $\square$

Thus almost uniform convergence implies almost everywhere convergence,

$$f_n \to f \text{ a.u.} \quad \overset{\Longrightarrow}{\not\Longleftarrow} \quad f_n \to f \text{ a.e.,}$$

but, in general, the converse fails. In fact, even pointwise convergence does not imply almost uniform convergence (see Problem 6.3). However, as we shall see in Proposition 6.13, the converse holds in a finite measure space.

Proposition 6.12. *Let $\{f_n\}$ be a sequence of measurable functions. If $f_n \to f$ almost uniformly, then $f_n \to f$ in measure.*

Proof. Take a measure space $(X, \mathcal{X}, \mu)$ and, for each $\alpha > 0$, put $F_n(\alpha) = \{x \in X \colon |f_n(x) - f(x)| \geq \alpha\}$ in $\mathcal{X}$. If $f_n \to f$ a.u., then for each $\delta > 0$ there exists a set E_δ in $\mathcal{X}$ with $\mu(E_\delta) < \delta$ and a positive integer $n_{\alpha,\delta}$ such that

$$n \geq n_{\alpha,\delta} \quad \text{implies} \quad \sup_{x \in X \setminus E_\delta} |f_n(x) - f(x)| < \alpha.$$

Hence $F_n(\alpha) \subseteq E_\delta$, and so $\mu(F_n(\alpha)) < \mu(E_\delta) < \delta$ for every $n \geq n_{\alpha,\delta}$. Therefore $\lim_n \mu(F_n(\alpha)) = 0$, which means that $f_n \to f$ in measure. $\qquad \square$

Thus almost uniform convergence also implies convergence in measure,

$$f_n \to f \text{ a.u.} \quad \underset{\Leftarrow}{\Rightarrow} \quad f_n \to f \text{ in measure,}$$

but the converse fails even if $\mu(X) < \infty$. Indeed, the sequence of Problem 6.2(b) acts on a finite measure space, converges in measure (since it converges in L^p), but does not converge a.u. (since it does not converge a.e.). Moreover, convergences almost uniform and in L^p are not related,

$$\{f_n\} \text{ converges a.u.} \quad \underset{\Leftarrow}{\Rightarrow} \quad \{f_n\} \text{ converges in } L^p,$$

even if $\mu(X) < \infty$. In fact, take the function $f_n = (n + 1) \mathcal{X}_{[1/(n+1),\, 2/(n+1)]}$ for each $n \geq 1$. It is clear that $\{f_n\}$ converges almost uniformly to the null function but it does not converge in L^p according to Problem 6.2(a). Conversely, we have just seen above that convergence in L^p does not imply almost uniform convergence. However, almost uniform dominated convergence implies convergence in L^p by Propositions 6.9 and 6.12.

Almost uniform convergence implies almost everywhere convergence (cf. Proposition 6.11). We close this chapter by showing in Proposition 6.13 that the converse holds in a finite measure space. This is referred to as the *Egoroff's Theorem*. Moreover, the finite measure assumption in Egoroff's Theorem can be replaced with dominated convergence (Corollary 6.14).

Proposition 6.13. *Let $\{f_n\}$ be a sequence of measurable functions. If $\mu(X) < \infty$ and $f_n \to f$ almost everywhere, then $f_n \to f$ almost uniformly.*

Proof. Consider a measure space $(X, \mathcal{X}, \mu)$ and suppose $\{f_n\}$ converges almost everywhere to f. Thus $f_n(x) \to f(x)$ for every x in $X \setminus N \in \mathcal{X}$ for some $N \in \mathcal{X}$ with $\mu(N) = 0$. That is, $f_n \to f$ pointwise on the complement $X' = X \backslash N$ of a set of measure zero or, simply, $f'_n \to f'$ pointwise on X with $f'_n = f_n \chi_{X'}$ for each n and $f' = f \chi_{X'}$. That is, $f'_n(x) \to f'(x)$ for every $x \in X$. Take an arbitrary positive integer m and set, for each $n \geq 1$,

$$F'_n(m) = \{x \in X' : |f_n(x) - f(x)| \geq \tfrac{1}{m}\} = \{x \in X : |f'_n(x) - f'(x)| \geq \tfrac{1}{m}\}.$$

Since f' and f'_n are measurable (because f and f_n are), each $|f'_n - f'|$ is a measurable function, and hence each $F'_n(m)$ is a measurable set. Then

$$E'_n(m) = \bigcup_{k=n}^{\infty} F'_k(m)$$

is a measurable set for each $n \geq 1$, and so $\{E'_n(m)\}$ is a decreasing sequence of sets in $\mathcal{X}$ (i.e., $E'_{n+1}(m) \subseteq E'_n(m) \in \mathcal{X}$). Next take an arbitrary x in X. Since $f'_n(x) \to f'(x)$, it follows that there exists an integer $n_{m,x}$ such that

$$k \geq n_{m,x} \implies |f'_k(x) - f'(x)| < \tfrac{1}{m} \implies x \notin F'_k(m) \implies x \notin E'_{n_{m,x}}(m),$$

and so $x \notin \bigcap_{n=1}^{\infty} E'_n(m)$. Hence

$$\bigcap_{n=1}^{\infty} E'_n(m) = \varnothing, \quad \text{which implies} \quad \mu\Big(\bigcap_{n=1}^{\infty} E'_n(m)\Big) = 0.$$

Now we use the finite measure assumption. According to Proposition 2.2(d),

$$\mu(X) < \infty \implies \mu(E'_1(m)) < \infty \implies \lim_n \mu(E'_n(m)) = 0.$$

Thus, for every $\varepsilon > 0$ there exists an integer $n_{\varepsilon,m}$ such that $\mu(E'_n(m)) < \tfrac{\varepsilon}{2^m}$ for every $n \geq n_{\varepsilon,m}$. Put $E'_\varepsilon = \bigcup_{m=1}^{\infty} E'_{n_{\varepsilon,m}}(m)$ in $\mathcal{X}$ so that

$$\mu(E'_\varepsilon) = \mu\Big(\bigcup_{m=1}^{\infty} E'_{n_{\varepsilon,m}}(m)\Big) \leq \sum_{m=1}^{\infty} \mu\big(E'_{n_{\varepsilon,m}}(m)\big) < \sum_{m=1}^{\infty} \frac{\varepsilon}{2^m} = \varepsilon.$$

Moreover, if $x \in X \backslash E'_\varepsilon$, then $x \notin E'_{n_{\varepsilon,m}}(m)$, so

$$n \geq n_{\varepsilon,m} \implies x \notin F'_n(m) \implies |f'_n(x) - f'(x)| < \tfrac{1}{m},$$

and $\{f'_n\}$ converges uniformly to f' on $X \backslash E'_\varepsilon$. Put $E_\varepsilon = E'_\varepsilon \cup N$ in $\mathcal{X}$. Since $X \backslash E_\varepsilon = X \backslash (E'_\varepsilon \cup N) = (X \backslash E'_\varepsilon) \cap (X \backslash N) = (X \backslash E'_\varepsilon) \cap X'$, we get

$$\sup_{x \in X \backslash E_\varepsilon} |f_n(x) - f(x)| = \sup_{x \in X \backslash E'_\varepsilon} |f'_n(x) - f'(x)|,$$

and so $\{f_n\}$ converges uniformly to f on $X \backslash E_\varepsilon$ as $\{f'_n\}$ converges uniformly to f' on $X \backslash E'_\varepsilon$. Therefore, since $\mu(E_\varepsilon) \leq \mu(E'_\varepsilon) + \mu(N) = \mu(E'_\varepsilon) < \varepsilon$, it follows that $\{f_n\}$ converges almost uniformly. $\qquad \square$

Corollary 6.14. *Let* $\{f_n\}$ *be a sequence of measurable functions. If* $|f_n| \le g \in L^p$ *and* $f_n \to f$ *almost everywhere, then* $f_n \to f$ *almost uniformly.*

Proof. Consider the proof of Proposition 6.13. The finite measure assumption was applied there only to ensure that $\mu(E_1'(m)) < \infty$. The present proof consists in showing that $E_1'(m)$ still has finite measure if we assume dominated convergence instead. Indeed, if $|f_n| \le g \in L^p$ and $f_n \to f$ a.e., then $|f| \le g$ (so that f_n and f lie in L^p) and $|f_n - f| \le |f_n| + |f| \le 2g$ (Proposition 6.4). Now, with $G'(m) = \{x \in X' : 2g(x) \ge \frac{1}{m}\}$ in X, we get

$$F_n'(m) = \left\{x \in X' : \tfrac{1}{m} \le |f_n(x) - f(x)|\right\} \subseteq \left\{x \in X' : \tfrac{1}{m} \le 2g(x)\right\} = G'(m)$$

for all n, and therefore

$$E_1'(m) = \bigcup_{n=1}^{\infty} F_n'(m) \subseteq G'(m).$$

Since $\int g^p \, d\mu < \infty$, it follows that $\mu(\{x \in X : g^p(x) \ge \varepsilon\}) < \infty$ for every $\varepsilon > 0$ (Problem 3.9), and hence $\mu(G'(m)) < \infty$. Thus $\mu(E_1'(m)) < \infty$. $\square$

6.4   Problems

Problem 6.1. Consider the Lebesgue measure space $(\mathbb{R}, \mathfrak{R}, \lambda)$.

(a) If $f_n = n^{-1/p} \mathcal{X}_{[0,\,n]}$ for each $n \ge 1$, then $\{f_n\}$ converges uniformly to the null function 0 but does not converge in L^p for any $p \ge 1$.

   Hint: If $n \ge 2m$, then show that $\|f_n - f_m\|_p^p \ge (n - m)m \ge m^2$.

(b) If $f_n = n^{1/p} \mathcal{X}_{[n,\,n+(1/n^2)]}$ for each $n \ge 1$, then $\{f_n\}$ converges in L^p to the null function 0 for every $p \ge 1$ but does not converge uniformly.

   Hint: $\{f_n\}$ converges pointwise to 0 but $\sup_{x \in \mathbb{R}} |f_n(x)| = n^{1/p}$.

Problem 6.2. Consider the *finite* measure space $\big([0,1], \wp([0,1]) \cap \mathfrak{R}, \lambda\big)$, where λ is the Lebesgue measure on $\wp([0,1]) \cap \mathfrak{R}$ (i.e., the restriction of the Lebesgue measure to the Borel subsets of $[0,1]$.)

(a) If $f_n = (n+1) \mathcal{X}_{[1/(n+1),\,2/(n+1)]}$ for $n \ge 1$, then $\{f_n\}$ converges pointwise to the null function 0 but does not converge in L^p for any $p \ge 1$.

   Hint: If $n \ge 2m + 1$, then $\|f_n - f_m\|_p^p \ge (2^{p-1} + 1)(m + 1)^{p-1}$.

   Note that $\{f_n\}$ does not converge uniformly (since it converges pointwise to 0 but $\sup_{x \in [0,1]} |f_n(x)| = (n+1)$), which in fact is a consequence of Proposition 6.3(b) once it does not converge in L^p.

(b) Set $E_{k,j} = \left[\frac{j-1}{k}, \frac{j}{k}\right]$ for each pair of integers j and k such that $1 \le j \le k$. For each $k \ge 1$ consider the finite sequence $\{E_{k,j}\}_{1 \le j \le k}$. Stack these finite sequences to get the infinite sequence of intervals

$$\{E_n\}_{n \ge 1} = \{\{E_{k,j}\}_{1 \le j \le k}\}_{k \ge 1}$$
$$= \{\{E_{1,1}\}, \{E_{2,1}, E_{2,2}\}, \{E_{3,1}, E_{3,2}, E_{3,3}\}, \{E_{4,1}, E_{4,2}, E_{4,3}, E_{4,4}\}, ...\}$$
$$= \{[0,1], [0,\tfrac{1}{2}], [\tfrac{1}{2},1], [0,\tfrac{1}{3}], [\tfrac{1}{3},\tfrac{2}{3}], [\tfrac{2}{3},1], [0,\tfrac{1}{4}], [\tfrac{1}{4},\tfrac{2}{4}], [\tfrac{2}{4},\tfrac{3}{4}], [\tfrac{3}{4},1], ...\}.$$

If $f_n = \chi_{E_n}$ for each $n \ge 1$, then $\{f_n\}$ converges in L^p to the null function 0 for every $p \ge 1$, but the real-valued sequence $\{f_n(x)\}$ *does not converge for every x in $[0,1]$* (i.e., $\{f_n\}$ does not converge pointwise everywhere, and so it does not converge almost everywhere).

Hint: The real-valued sequence $\{\lambda(E_n)\}$ is bounded and decreasing, thus convergent. For any $m \ge 1$ there is an n_m such that $\lambda(E_{n_m}) \le \frac{1}{m}$. Hence $\|f_n\|_p \to 0$. Now take an arbitrary x in $[0,1]$. The real-valued sequence $\{f_n(x)\}$ has a subsequence constantly equal to 1 and another constantly equal to 0. Hence $\{f_n(x)\}$ does not converge.

Problem 6.3. Take the Lebesgue measure space $(\mathbb{R}, \mathfrak{R}, \lambda)$ and, for each $n \ge 1$, consider the characteristic function $f_n = \chi_{[n, n+1]}$.

(a) $\{f_n\}$ converges pointwise (and so a.e.) to the null function 0.

(b) $\{f_n\}$ does not converge in measure (so not uniformly and not in L^p).

Hint: $\mu(\{x \in X: |f_m(x) - f_n(x)| \ge \frac{1}{2}\}) = 2$ whenever $m \ne n$, and hence $\{f_n\}$ is not Cauchy in measure (Proposition 6.6(a)).

(c) $\{f_n\}$ does not converge almost uniformly (and so not uniformly a.e.).

Problem 6.4. Consider the following real-valued functions on $\mathbb{R}$.

$$g(x) = \begin{cases} 0, & x \le 0, \\ \frac{1}{\sqrt{x}}, & x \in [0,1], \\ \frac{1}{x^2}, & x \in [1,\infty), \end{cases} \quad \text{and} \quad f_n(x) = \begin{cases} n, & g(x) \ge n, \\ 0, & \text{otherwise.} \end{cases}$$

Take the Lebesgue measure space $(\mathbb{R}, \mathfrak{R}, \lambda)$. Show: (i) $\{f_n\}$ is dominated by $g \in L^1$ and converges pointwise to the null function; (ii) $\{f_n\}$ converges to 0 almost uniformly but does not converge uniformly almost everywhere.

Hint: $\sup_{x \in X \setminus N} |f_n(x)| \to \infty$ for every set N of Lebesgue measure zero.

Problem 6.5. The *symmetric difference* of two sets A and B is the set

$$A \triangledown B = (A \setminus B) \cup (B \setminus A) = (A \cup B) \setminus (A \cap B).$$

Let $(X, \mathcal{X}, \mu)$ be a measure space and take arbitrary sets E and F in $\mathcal{X}$. We say that E and F are *equivalent* (or *μ-equivalent*), denoted by $E \sim F$, if $\mu(E \triangledown F) = 0$. This $\sim$ is indeed an equivalence relation on $\mathcal{X}$. Define the function $d: \mathcal{X} \times \mathcal{X} \to \mathbb{R}$ by $d(E, F) = \mu(E \triangledown F)$ for every E, F in $\mathcal{X}$. Verify that $d(E, F) \geq 0$, $d(E, F) = d(F, E)$, and $d(E, F) \leq d(E, G) + d(G, F)$ for every E, F, G in $\mathcal{X}$. (Note: d is then a *pseudometric* on $\mathcal{X}$, and so it induces a *metric* on the quotient space $\mathcal{X}/\sim$.) Now take a sequence $\{E_n\}$ of sets in $\mathcal{X}$ and, for each $n \geq 1$, consider the characteristic function $f_n = \chi_{E_n}$. Show that $\{f_n\}$ is Cauchy in measure if and only if $\lim_{m,n} d(E_m, E_n) = 0$.

Problem 6.6. The next diagrams show the relationship among almost everywhere convergence, almost uniform convergence, convergence in L^p, and convergence in measure, denoted by (a.e.), (a.u.), (L^p), and (μ), respectively. The first diagram considers the general case (no additional assumption), the second one considers the case of finite measure $(\mu(X) < \infty)$, and the third diagram considers the case of dominated convergence $(|f_n| \leq g \in L^p)$.

(a.e.) $\Longleftarrow$ (a.u.)	(a.e.) $\Longleftrightarrow$ (a.u.)	(a.e.) $\Longleftrightarrow$ (a.u.)
$\Big\Downarrow$	$\Big\Downarrow$	$\Big\Downarrow$
$(L^p) \Longrightarrow (\mu)$	$(L^p) \Longrightarrow (\mu)$	$(L^p) \Longleftrightarrow (\mu)$
General case	*Finite measure*	*Dominated convergence*

Clearly, the arrows mean implication. Verify that those diagrams are correct and complete, in the sense that all arrows are true and no arrow can be added, except the trivial ones (i.e., up to *modus ponens* — for instance, it is obvious from those diagrams that (a.e.) implies (μ) in a finite measure space and also that (a.e.) implies (L^p) under the dominance assumption).

Problem 6.7. Let (u.a.e.) denote uniform convergence almost everywhere. Verify the following implications and also that their converses fail even under finite measure and dominance assumptions.

$$\text{(u.a.e.)} \implies \text{(a.e.)} \qquad \text{and} \qquad \text{(u.a.e.)} \implies \text{(a.u.)}$$

Show that (u.a.e.) implies (L^p) under finite measure or dominance but not in general and that the converse fails even under both assumptions.

Suggested Reading

Bartle [3], Berberian [6], Halmos [13], Munroe [19].

7

Decomposition

7.1 Hahn and Jordan Decompositions

Let $\nu: \mathcal{X} \to \mathbb{R}$ be a signed measure on a σ-algebra $\mathcal{X}$ of subsets of a set X. As we have defined them (cf. Definition 2.3), signed measures are *real-valued* set functions. We saw in Section 2.2 that if μ and λ are finite measures, then $\nu = \mu - \lambda$ is a signed measure. In this section we show that all signed measures admit a decomposition into a difference of finite measures.

Definition 7.1. Let ν be a signed measure on a σ-algebra $\mathcal{X}$. A set A^+ in $\mathcal{X}$ is *positive* with respect to ν if $\nu(A^+ \cap E) \geq 0$ for all E in $\mathcal{X}$. A set A^- in $\mathcal{X}$ is *negative* with respect to ν if $\nu(A^- \cap E) \leq 0$ for all E in $\mathcal{X}$. A set N in $\mathcal{X}$ is *null* with respect to ν if $\nu(N \cap E) = 0$ for all E.

Equivalently, a measurable set is *positive, negative,* or *null* if each measurable subset of it has nonnegative, nonpositive, or null measure, respectively. Every set X has a measurable partition into a positive and a negative set with respect to any signed measure ν on any σ-algebra $\mathcal{X}$ of subsets of X.

Theorem 7.2. (Hahn Decomposition Theorem). *If ν is a signed measure on $\mathcal{X}$, then there exists a measurable partition $\{A^+, A^-\}$ of X such that A^+ is positive and A^- is negative with respect to ν.*

Proof. Let ν be a signed measure on a σ-algebra $\mathcal{X}$ of subsets of a set X. We prove the existence of a pair of sets A^+ and A^- in $\mathcal{X}$ such that $A^+ \cup A^- = X$, $A^+ \cap A^- = \emptyset$, A^+ is positive, and A^- is negative. Let $\mathcal{A} \subseteq \mathcal{X}$ be the collection of all positive sets with respect to ν, which is not empty (it contains the empty set). Put $\alpha = \sup_{A \in \mathcal{A}} \nu(A)$ and take a sequence $\{A_n\}$ of sets in $\mathcal{A}$ such that $\sup_n \nu(A_n) = \alpha$. Thus $A^+ = \bigcup_n A_n$ is a positive set (Problem 7.2(b)) with $0 \leq \nu(A^+) = \alpha < \infty$ (since $\nu(A_n) \leq \nu(A^+) \leq \alpha$ for all n; cf. Problem 7.3). If $A^- = X \backslash A^+$ is a negative set, then we are done.

Claim. $A^- = X \backslash A^+$ is a negative set.

Proof. If A^- is not negative, then it has a measurable subset E_0 such that $\nu(E_0) > 0$. If E_0 is a positive set, then $\nu(A^+ \cup E_0) > \alpha$ (for $A^+ \cap E_0 = \varnothing$), which is a contradiction ($\alpha = \sup_{A \in \mathcal{A}} \nu(A)$). Thus E_0 is not positive, so it has measurable subsets of negative measure. Let n_0 be the smallest positive integer such that E_0 has a measurable subset of measure not greater than $-\frac{1}{n_0}$, say E_1 with $\nu(E_1) \leq -\frac{1}{n_0}$. Note that

$$\nu(E_0 \backslash E_1) = \nu(E_0) - \nu(E_1) > \nu(E_0) > 0$$

(Problem 7.1(a)). If $E_0 \backslash E_1$ is a positive set, then $\nu(A^+ \cup (E_0 \backslash E_1)) > \alpha$ (for $A^+ \cap (E_0 \backslash E_1) = \varnothing$), which is again a contradiction. Thus $E_0 \backslash E_1$ is not positive, so it has measurable subsets of negative measure. Let n_1 be the smallest positive integer such that $E_0 \backslash E_1$ has a measurable subset of measure not greater than $-\frac{1}{n_1}$, say E_2 with $\nu(E_2) \leq -\frac{1}{n_1}$. Note that

$$\nu(E_0 \backslash (E_1 \cup E_2)) = \nu(E_0) - \nu(E_1 \cup E_2) = \nu(E_0) - (\nu(E_1) + \nu(E_2)) > \nu(E_0) > 0$$

(since $E_1 \cap E_2 = \varnothing$). As before, $E_0 \backslash (E_1 \cup E_2)$ is not positive, so it has measurable subsets of negative measure. Let n_2 be the smallest positive integer such that $E_0 \backslash (E_1 \cup E_2)$ has a measurable subset of measure not greater than $-\frac{1}{n_2}$, say E_3 with $\nu(E_3) \leq -\frac{1}{n_2}$. This leads to the inductive construction of a sequence $\{E_k\}_{k=1}^{\infty}$ of pairwise disjoint measurable sets and a sequence $\{n_k\}_{k=1}^{\infty}$ with each n_k being the smallest positive integer for which $E_0 \backslash \bigcup_{i=1}^{k} E_i$ has a measurable subset of measure not greater than $-\frac{1}{n_k}$. Moreover, $\nu(E_{k+1}) \leq -\frac{1}{n_k}$ for every $k \geq 0$, and so $\sum_{k=0}^{\infty} \frac{1}{n_k} < \infty$. Indeed, by setting $E = \bigcup_{k=1}^{\infty} E_k$ in X we get

$$-\infty < \nu(E) = \sum_{k=1}^{\infty} \nu(E_k) \leq -\sum_{k=0}^{\infty} \frac{1}{n_k} < 0$$

($\{E_k\}_{k=1}^{\infty}$ consists of disjoint sets). Hence $\frac{1}{n_k} \to 0$ as $k \to \infty$. Note that

$$\nu(E_0 \backslash E) = \nu(E_0) - \nu(E) > \nu(E_0) > 0.$$

$E_0 \backslash E$ actually is a positive set. In fact, suppose $E_0 \backslash E$ has a measurable subset of negative measure, say F with $\nu(F) < 0$. Since $n_k \to \infty$ as $k \to \infty$, take k large enough so that $\frac{1}{n_k - 1} < -\nu(F)$; that is, $\nu(F) < -\frac{1}{n_k - 1}$. But

$F \subseteq E_0 \backslash E \subseteq E_0 \backslash \bigcup_{i=1}^k E_i$, so $E_0 \backslash \bigcup_{i=1}^k E_i$ has a measurable subset of measure less than $-\frac{1}{n_k - 1}$, which contradicts the fact that n_k is the smallest positive integer for which $E_0 \backslash \bigcup_{i=1}^k E_i$ has a measurable subset of measure not greater than $-\frac{1}{n_k}$. Thus every measurable subset of $E_0 \backslash E$ has a nonnegative measure, and so $E_0 \backslash E$ is a positive set. Hence, since $A^+ \cap (E_0 \backslash E) = \varnothing$ (because $E_0 \subseteq A^-$) and $\nu(E_0 \backslash E) > 0$, it follows that $\nu(A^+ \cup (E_0 \backslash E)) > \alpha$, which is a contradiction. Therefore, A^- must be a negative set.   □

A measurable partition $\{A^+, A^-\}$ of X, where A^+ is positive and A^- is negative with respect to a signed measure ν on $\mathcal{X}$, is called a *Hahn decomposition* of X with respect to the ν. Given a signed measure ν on $\mathcal{X}$, a Hahn decomposition of X is not necessarily unique. Indeed, if $\{A^+, A^-\}$ is a Hahn decomposition of X and N is a null set, then $\{A^+ \cup N, A^- \backslash N\}$ and $\{A^+ \backslash N, A^- \cup N\}$ are Hahn decompositions of X (all with respect to ν). This lack of uniqueness, however, is transparent for the signed measure ν and hence is not a drawback as far as most applications are concerned.

Proposition 7.3. *If* $\{A_1^+, A_1^-\}$ *and* $\{A_2^+, A_2^-\}$ *are Hahn decompositions of* X *with respect to a signed measure* ν *on* $\mathcal{X}$ *then, for every* $E \in \mathcal{X}$,

$$\nu(A_1^+ \cap E) = \nu(A_2^+ \cap E) \quad \text{and} \quad \nu(A_1^- \cap E) = \nu(A_2^- \cap E).$$

Proof. If A, B, and C are sets, then $\{A \cap (B \backslash C), A \cap B \cap C\}$ is a partition of $A \cap B$. In $\mathcal{X}$, if $\nu(A \cap (B \backslash C)) = 0$, then $\nu(A \cap B) = \nu(A \cap B \cap C)$. Since $E \cap (A_1^+ \backslash A_2^+) \subseteq A_1^+ \cap A_2^-$ and $E \cap (A_2^+ \backslash A_1^+) \subseteq A_2^+ \cap A_1^-$, it follows that

$$\nu(E \cap (A_1^+ \backslash A_2^+)) = 0 \quad \text{and} \quad \nu(E \cap (A_2^+ \backslash A_1^+)) = 0.$$

Hence

$$\nu(E \cap A_1^+) = \nu(E \cap A_1^+ \cap A_2^+) \quad \text{and} \quad \nu(E \cap A_2^+) = \nu(E \cap A_2^+ \cap A_1^+),$$

and therefore

$$\nu(E \cap A_1^+) = \nu(E \cap A_2^+).$$

Similarly, swapping A_1^+ and A_1^-, and also A_2^+ and A_2^-, we get

$$\nu(E \cap A_1^-) = \nu(E \cap A_2^-).$$ □

Let $\{A^+, A^-\}$ be a Hahn decomposition of X with respect to a signed measure ν on $\mathcal{X}$. It is readily verified (cf. Problem 2.11) that the functions $\nu^+ : \mathcal{X} \to \mathbb{R}$ and $\nu^- : \mathcal{X} \to \mathbb{R}$ defined, for every set E in $\mathcal{X}$, by

$$\nu^+(E) = \nu(A^+ \cap E) \quad \text{and} \quad \nu^-(E) = -\nu(A^- \cap E)$$

are finite measures on $\mathcal{X}$. These measures ν^+ and ν^- are referred to as the *positive variation* and the *negative variation* of ν, respectively, which are unambiguously defined (their definitions do not depend on the Hahn decomposition $\{A^+, A^-\}$ according to Proposition 7.3).

Theorem 7.4. (Jordan Decomposition Theorem). *For a signed measure ν,*

$$\nu = \nu^+ - \nu^-.$$

Moreover, if a signed measure ν on $\mathcal{X}$ is such that

$$\nu = \lambda - \mu,$$

where λ and μ are finite measures on $\mathcal{X}$, then

$$\nu^+ \le \lambda \quad \text{and} \quad \nu^- \le \mu.$$

Proof. Let $\{A^+, A^-\}$ be a Hahn decomposition of X with respect to a signed measure ν on a σ-algebra $\mathcal{X}$ of subsets of X. Since $\{A^+, A^-\}$ is a partition of X, $(A^+ \cap E) \cup (A^- \cap E) = E$ and $(A^+ \cap E) \cap (A^- \cap E) = \varnothing$. Thus

$$\nu(E) = \nu(A^+ \cap E) + \nu(A^- \cap E) = \nu^+(E) - \nu^-(E)$$

for every $E \in \mathcal{X}$. If λ and μ are finite measures on $\mathcal{X}$ (so they have a finite nonnegative value for each measurable set) such that $\nu = \lambda - \mu$, then

$$\nu^+(E) = \nu(A^+ \cap E) = \lambda(A^+ \cap E) - \mu(A^+ \cap E) \le \lambda(A^+ \cap E) \le \lambda(E),$$
$$\nu^-(E) = -\nu(A^- \cap E) = -\lambda(A^- \cap E) + \mu(A^+ \cap E) \le \mu(A^+ \cap E) \le \mu(E),$$

for every $E \in \mathcal{X}$ (cf. Proposition 2.2(a)). That is, $\nu^+ \le \lambda$ and $\nu^- \le \mu$. $\square$

The sum of finite measures is again a finite measure. The *total variation* of a signed measure $\nu \colon \mathcal{X} \to \mathbb{R}$ is the finite measure $|\nu| \colon \mathcal{X} \to \mathbb{R}$ defined by

$$|\nu| = \nu^+ + \nu^-.$$

Example 7A. If ν is a signed measure on $\mathcal{X}$, then the total variation $|\nu|$ coincides with the (plain) variation μ defined in Example 2I. In fact, we have seen in Example 2I that

$$\mu(E) = \sup_{\{E^+, E^-\} \in \boldsymbol{E}(2)} \big(\nu(E^+) - \nu(E^-)\big) \quad \text{for every} \quad E \in \mathcal{X},$$

where the supremum is taken over all measurable partitions $\{E^+, E^-\}$ of E consisting of two sets such that $\nu(E^+) \ge 0$ and $\nu(E^-) \le 0$. If $\{A^+, A^-\}$ is any Hahn decomposition of X with respect to ν and $\mathcal{E} = \wp(E) \cap \mathcal{X}$, then

$$\nu^+(E) + \nu^-(E) = \nu(A^+ \cap E) - \nu(A^- \cap E) \le \mu(E)$$
$$\le \sup_{F \in \mathcal{E}} \nu(F) - \inf_{F \in \mathcal{E}} \nu(F) = \nu^+(E) + \nu^-(E),$$

where the last identity follows from Theorem 7.4 (via Problem 7.5), and so

$$\mu(E) = |\nu|(E) = \nu^+(E) + \nu^-(E) \quad \text{for every} \quad E \in \mathcal{X}.$$

Proposition 7.5. *Let* $(X, \mathcal{X}, \mu)$ *be a measure space. Take* $f \in \mathcal{L}(X, \mathcal{X}, \mu)$. *If* $\nu : \mathcal{X} \to \mathbb{R}$ *is the signed measure defined in* Lemma 4.6, *that is,*

$$\nu(E) = \int_E f \, d\mu \quad \text{for every} \quad E \in \mathcal{X},$$

then the negative, positive, and total variations of ν *are given for* $E \in \mathcal{X}$ *by*

$$\nu^+(E) = \int_E f^+ d\mu, \quad \nu^-(E) = \int_E f^- d\mu, \quad \text{and} \quad |\nu|(E) = \int_E |f| d\mu.$$

Proof. Consider sets $F_+ = \{x \in X : f(x) > 0\}$, $F_- = \{x \in X : f(x) < 0\}$, and $F_0 = \{x \in X : f(x) = 0\}$, and put

$$F^+ = F_+ \cup F_0 = \{x \in X : f \ge 0\} \quad \text{and} \quad F^- = F_- \cup F_0 = \{x \in X : f \le 0\}.$$

$\{F^+, F_-\}$ is a measurable partition of X such that $\nu(F^+ \cap E) \ge 0$ and $\nu(F_- \cap E) \le 0$ for each $E \in \mathcal{X}$, and so $\{F^+, F_-\}$ is a Hahn decomposition of X with respect to ν. Since $\int_{F_- \cap E} f \, d\mu = \int_{F^- \cap E} f \, d\mu$ (cf. Proposition 3.7(a), Problem 3.8(a), and Definition 4.1), it follows for every $E \in \mathcal{X}$ that

$$\nu^+(E) = \nu(F^+ \cap E) = \int_{F^+ \cap E} f \, d\mu = \int_E f \chi_{F^+} \, d\mu = \int_E f^+ d\mu,$$

$$\nu^-(E) = -\nu(F_- \cap E) = -\int_{F_- \cap E} f \, d\mu = \int_E -f \chi_{F^-} \, d\mu = \int_E f^- d\mu,$$

$$|\nu|(E) = \nu^+(E) + \nu^-(E) = \int_E (f^+ + f^-) \, d\mu = \int_E |f| \, d\mu. \qquad \square$$

7.2   The Radon–Nikodým Theorem

Definition 7.6. Let $(X, \mathcal{X})$ be a measurable space. A measure λ on $\mathcal{X}$ is *absolutely continuous* with respect to a measure μ on $\mathcal{X}$ if, for $E \in \mathcal{X}$,

$$\mu(E) = 0 \quad \text{implies} \quad \lambda(E) = 0$$

(i.e., $\lambda(E) = 0$ for every $E \in \mathcal{X}$ such that $\mu(E) = 0$). Notation: $\lambda \ll \mu$.

Let λ and μ be measures on $\mathcal{X}$ and consider the following assertions.

(a) For every $\varepsilon > 0$ there exists $\delta_\varepsilon > 0$ such that, for $E \in \mathcal{X}$,

$$\mu(E) < \delta_\varepsilon \quad \text{implies} \quad \lambda(E) < \varepsilon.$$

(b) λ is absolutely continuous with respect to μ (i.e., $\lambda \ll \mu$).

We see in Proposition 7.7 that these assertions are equivalent if λ is a finite measure, which justifies the terminology "absolute continuity".

Proposition 7.7. (a) *implies* (b), *and* (b) *implies* (a) *if* λ *is finite.*

Proof. If (a) holds true and $\mu(E) = 0$ for some $E \in \mathcal{X}$, then $\lambda(E) < \varepsilon$ for all $\varepsilon > 0$, which means that $\lambda(E) = 0$, and therefore (a) implies (b). Conversely, if (a) fails, then there exists an $\varepsilon > 0$ such that for every $\delta > 0$ there exists $E_\delta \in \mathcal{X}$ for which $\mu(E_\delta) < \delta$ and $\lambda(E_\delta) \geq \varepsilon$. In particular, for each $n \geq 1$ there exists a set E_n in $\mathcal{X}$ such that $\mu(E_n) < \frac{1}{2^n}$ and $\lambda(E_n) \geq \varepsilon$. Put $F_n = \bigcup_{k=n}^{\infty} E_k$ in $\mathcal{X}$ so that $\mu(F_n) \leq \sum_{k=n}^{\infty} \mu(E_k) < \sum_{k=n}^{\infty} \frac{1}{2^k} = \frac{1}{2^{n-1}}$ and $\lambda(F_n) \geq \lambda(E_k) \geq \varepsilon$ for every $n \geq 1$. Now set $F = \bigcap_{n=1}^{\infty} F_n$ in $\mathcal{X}$. Since $\{F_n\}$ is a decreasing sequence of sets in $\mathcal{X}$, and since $\mu(F_1) \leq 1$ and $\lambda(F_1) < \infty$ if λ is a finite measure, it follows by Proposition 2.2(d) that

$$\mu(F) = \lim_n \mu(F_n) = 0 \quad \text{and} \quad \lambda(F) = \lim_n \lambda(F_n) \geq \varepsilon,$$

and so (b) fails. Equivalently, (b) implies (a) if λ is a finite measure. $\square$

Recall from Propositions 3.5(c) and 3.7(b): *if* μ *is a measure on* $\mathcal{X}$ *and* f *is a function in* $\mathcal{M}(X, \mathcal{X})^+$ (nonnegative extended real-valued measurable function), *then the set function* λ *on* $\mathcal{X}$, *defined for each* $E \in \mathcal{X}$ *by*

$$\lambda(E) = \int_E f \, d\mu,$$

is a measure that is absolutely continuous with respect to μ. What might come as a nice, perhaps unexpected, result is that the converse holds if λ and μ are σ-finite. That is, in such a case, there exists a function f in $\mathcal{M}(X, \mathcal{X})^+$ for which λ is expressed as an integral of f with respect to μ. Moreover, such a function is unique μ-almost everywhere; that is, if g in $\mathcal{M}(X, \mathcal{X})^+$ is such that $\lambda(E) = \int_E g \, d\mu$ for every $E \in \mathcal{X}$, then $g = f$ μ-a.e. This is a central result in measure theory, which we see next.

Theorem 7.8. (Radon–Nikodým Theorem). *Let* $(X, \mathcal{X})$ *be a measurable space. If* λ *and* μ *are* σ-*finite measures on* $\mathcal{X}$, *and if* λ *is absolutely continuous with respect to* μ, *then there exists a unique* (μ-*almost everywhere unique*) *real-valued function* f *in* $\mathcal{M}(X, \mathcal{X})^+$ *such that, for each* $E \in \mathcal{X}$,

$$\lambda(E) = \int_E f \, d\mu.$$

Proof. Let λ and μ be measures on $\mathcal{X}$ such that $\lambda \ll \mu$. We split the proof into two parts. The theorem is proved for finite measures in part (a). This will be extended to σ-finite measures in part (b).

(a) Take an arbitrary positive number α. If λ and μ are finite measures, then consider the (real-valued) signed measure $\nu_\alpha = \lambda - \alpha\mu$ and let $\{A_\alpha^+, A_\alpha^-\}$ be a Hahn decomposition for X with respect to ν_α. Consider a sequence $\{E_k\}_{k\geq 1}$ of sets in $\mathcal{X}$ recursively defined by

$$E_{k+1} = A_{(k+1)\alpha}^- \Big\backslash \bigcup_{j=1}^k E_j, \qquad E_1 = A_\alpha^-.$$

It is readily verified by induction that

(i) $\{E_k\}_{k\geq 1}$ is a sequence of disjoint sets,

(ii) $\displaystyle \bigcup_{j=1}^k E_j = \bigcup_{j=1}^k A_{j\alpha}^-$ for every $k \geq 1$,

and so

$$E_k = A_{k\alpha}^- \Big\backslash \bigcup_{j=1}^{k-1} A_{j\alpha}^- = A_{k\alpha}^- \cap \bigcap_{j=1}^{k-1} A_{j\alpha}^+ \quad \text{for every } k \geq 2.$$

Thus $E_k \in (A_{k\alpha}^- \cap A_{(k-1)\alpha}^+)$, which implies, for any set $E \subseteq (\mathcal{X} \cap E_k)$, that $\lambda(E) - k\alpha\mu(E) \leq 0$ and $\lambda(E) - (k-1)\alpha\mu(E) \geq 0$, and hence

(iii) $(k-1)\alpha\mu(E) \leq \lambda(E) \leq k\alpha\mu(E),$

for every $E \subseteq (\mathcal{X} \cap E_k)$, for each $k \geq 2$. Now put (cf. property (ii))

$$F = X \Big\backslash \bigcup_{j=1}^\infty E_j = X \Big\backslash \bigcup_{j=1}^\infty A_{j\alpha}^- = \bigcap_{j=1}^\infty A_{j\alpha}^+ \subseteq A_{k\alpha}^+ \quad \text{for all } k \geq 1$$

so that $\lambda(F) - k\alpha\mu(F) \geq 0$, that is, $0 \leq k\alpha\mu(F) \leq \lambda(F)$, for all $k \geq 1$. Since λ is a finite measure, it follows that $\mu(F) = 0$, and therefore

(iv) $\lambda(F) = 0$

because $\lambda \ll \mu$. Next consider the nonnegative real-valued function f_α in $\mathcal{M}^+ = \mathcal{M}(X, \mathcal{X})^+$ defined by

$$f_\alpha(x) = \begin{cases} (k-1)\alpha, & x \in E_k \text{ for some } k \geq 1, \\ 0, & x \in F = X \backslash \bigcup_{k=1}^\infty E_k, \end{cases}$$

and recall that $\{F, E_k; k \geq 1\}$ is measurable partition of X by property (i). Take an arbitrary $E \in \mathcal{X}$ so that $\{E \cap F, E \cap E_k; k \geq 1\}$ is a measurable partition of E. Thus (cf. Problems 3.3(a) and 4.10)

$$\int_E f_\alpha \, d\mu = \int_{\bigcup_{k=1}^\infty (E \cap E_k)} f_\alpha \, d\mu = \sum_{k=1}^\infty \int_{E \cap E_k} f_\alpha \, d\mu$$

$$= \sum_{k=1}^\infty (k-1)\alpha \int_{E \cap E_k} d\mu = \sum_{k=2}^\infty (k-1)\alpha \mu(E \cap E_k)$$

$$\leq \sum_{k=2}^\infty \lambda(E \cap E_k) = \lambda(\textstyle\bigcup_{k=2}^\infty (E \cap E_n)) \leq \lambda(E)$$

according to properties (iii) and (i). Similarly, applying properties (i), (iii), and (iv) we also get

$$\lambda(E) = \int_E d\lambda = \int_{\bigcup_{k=1}^\infty (E \cap E_k)} d\lambda = \lambda(\textstyle\bigcup_{k=1}^\infty (E \cap E_k)) = \sum_{k=1}^\infty \lambda(E \cap E_k)$$

$$\leq \sum_{k=1}^\infty k\alpha \mu(E \cap E_k) = \sum_{k=1}^\infty \int_{E \cap E_k} k\alpha \, d\mu = \sum_{k=1}^\infty \int_{E \cap E_k} (f_\alpha + \alpha) \, d\mu$$

$$= \int_{\bigcup_{k=1}^\infty (E \cap E_k)} (f_\alpha + \alpha) \, d\mu \leq \int_E (f_\alpha + \alpha) \, d\mu = \int_E f_\alpha \, d\mu + \alpha\mu(E).$$

Now take an arbitrary integer $n \geq 1$, set $\alpha = \left(\frac{1}{2}\right)^n$, and put $f_n = f_{(\frac{1}{2})^n}$. The previously displayed inequalities say that

(v) $$\int_E f_n \, d\mu \leq \lambda(E) \leq \int_E f_n \, d\mu + \left(\tfrac{1}{2}\right)^n \mu(X)$$

for all $n \geq 1$. Thus, for any pair of positive integers m and n,

$$\int_E f_n \, d\mu \leq \lambda(E) \leq \int_E f_m \, d\mu + \left(\tfrac{1}{2}\right)^m \mu(X),$$

$$\int_E f_m \, d\mu \leq \lambda(E) \leq \int_E f_n \, d\mu + \left(\tfrac{1}{2}\right)^n \mu(X),$$

and hence

$$\int_E f_n \, d\mu - \int_E f_m \, d\mu \leq \lambda(E) - \int_E f_m \, d\mu \leq \left(\tfrac{1}{2}\right)^m \mu(X),$$

$$\int_E f_m \, d\mu - \int_E f_n \, d\mu \leq \lambda(E) - \int_E f_n \, d\mu \leq \left(\tfrac{1}{2}\right)^n \mu(X),$$

so

$$\left| \int_E (f_m - f_n) \, d\mu \right| \leq \left(\tfrac{1}{2}\right)^m \mu(X)$$

whenever $m \leq n$. Since this holds for all $E \in \mathcal{X}$, we get

$$\int |f_m - f_n| \, d\mu = \int (f_m - f_n)^+ \, d\mu + \int (f_m - f_n)^- \, d\mu$$

$$= \int_{F_{m,n}^+} (f_m - f_n) \, d\mu - \int_{F_{m,n}^-} (f_m - f_n) \, d\mu$$

$$= \left| \int_{F_{m,n}^+} (f_m - f_n) \, d\mu \right| + \left| \int_{F_{m,n}^-} (f_m - f_n) \, d\mu \right| \leq \left(\tfrac{1}{2} \right)^m \mu(X)$$

for every m, n such that $m \leq n$, with $F_{m,n}^+ = \{x \in X \colon (f_m - f_n)(x) \geq 0\}$ and $F_{m,n}^- = \{x \in X \colon (f_m - f_n)(x) \leq 0\}$. Since property (v) holds for all E in $\mathcal{X}$, λ is a finite measure, and f_n is a real-valued function in $\mathcal{M}^+$, it follows that each f_n is in $L^1(\mu) = L^1(X, \mathcal{X}, \mu)$. Hence, by the above inequality,

$$\|f_m - f_n\|_1 \leq \left(\tfrac{1}{2} \right)^m \mu(X)$$

whenever $m \leq n$. Since μ also is a finite measure, the preceding inequality ensures that $\{f_n\}$ is a Cauchy sequence in the Banach space $L^1(\mu)$, and so it converges in $L^1(\mu)$ to, say, $f \in L^1(\mu)$. This implies that the real-valued sequence $\{\int_E f_n \, d\mu\}$ converges in $\mathbb{R}$ to $\int_E f \, d\mu$ for every $E \in \mathcal{X}$. Indeed,

$$\left| \int_E f_n \, d\mu - \int_E f \, d\mu \right| \leq \int_E |f_n - f| \, d\mu \leq \int |f_n - f| \, d\mu = \|f_n - f\|_1 \to 0.$$

(Note that $f = \lim_n f_n$ in $L^1(\mu)$ is real-valued.) Therefore, by property (v),

$$\lambda(E) = \lim_n \int_E f_n \, d\mu = \int_E f \, d\mu \quad \text{for every } E \in \mathcal{X}.$$

Observe that we may take a nonnegative function f in the equivalence class $[f]$, that is, we may take $f \in \mathcal{M}^+$. In fact, since $0 \leq \lambda(E) = \int_E f \, d\mu$ for every $E \in \mathcal{X}$, it follows by Problem 4.5(b) that $f \geq 0$ μ-a.e. (and also λ-a.e. because $\lambda \ll \mu$) for all $f \in [f]$. Moreover, this function f is unique μ-a.e. Indeed, if $g \in \mathcal{M}^+$ (or $g \in L^1(\mu)$) is such that $\lambda(E) = \int_E f \, d\mu = \int_E g \, d\mu$ for every $E \in \mathcal{X}$, then $f = g$ μ-a.e. by Problem 3.8(d) (or Problem 4.5(c)).

(b) Now suppose the measures λ and μ are σ-finite so that there exists a pair of sequences of $\mathcal{X}$-measurable sets, say $\{A_n\}$ and $\{B_n\}$, such that $\lambda(A_n) < \infty$ and $\mu(B_n) < \infty$ for every n, both covering X. Suppose these sequences are increasing, which does not imply any loss of generality (why?). Put $X_n = A_n \cap B_n$ so that $\{X_n\}$ is an increasing sequence of $\mathcal{X}$-measurable sets such that $\bigcup X_n = \bigcup A_n \cap \bigcup B_n = X$ and, for each n,

$$\lambda(X_n) < \infty \quad \text{and} \quad \mu(X_n) < \infty.$$

Take an arbitrary index n and consider the σ-algebra $\mathcal{X}_n = \mathcal{P}(X_n) \cap \mathcal{X}$. According to part (a) there exists a function g_n in $\mathcal{M}(X_n, \mathcal{X}_n)^+$ such that $\lambda(E') = \int_{E'} g_n \, d\mu|_{X_n}$ for every $E' \in \mathcal{X}_n$ (see Problem 2.11). Thus the function f_n defined by $f_n(x) = g_n(x)$ for $x \in X_n$ and $f_n(x) = 0$ for $x \in X \backslash X_n$ (i.e., $f_n = g_n \chi_{X_n}$) lies in $\mathcal{M}^+ = \mathcal{M}(X, \mathcal{X})^+$ and, for every $E' \in \mathcal{X}_n$,

$$\lambda(E') = \int_{E'} f_n \, d\mu.$$

Since $X_n \subseteq X_k$ for every $k \geq n$, it follows that $E' \in \mathcal{X}_k$ for every $k \geq n$ if $E' \in \mathcal{X}_n$, and so the above identity holds for all $k \geq n$. That is, the sequence $\{f_n\}$ of functions in $\mathcal{M}^+$ is such that, if $E' \in \mathcal{X}_n$ for some $n \geq 1$, then

$$\lambda(E') = \int_{E'} f_n \, d\mu = \int_{E'} f_k \, d\mu$$

for all $k \geq n$. From part (a) we know that g_n is unique μ-a.e., then so is f_n, for each n. Thus the foregoing identity ensures that $f_k = f_n$ μ-a.e. on X_n for all $k \geq n$ (same uniqueness argument: apply Problem 3.8(d) or Problem 4.5(a) — nonnegative functions in $L^1(\mu)$). Therefore, since each f_n vanishes outside X_n, and since $\{X_n\}$ is an increasing sequence of sets, it follows that $\{f_n\}$ is an increasing sequence of functions in $\mathcal{M}^+$ such that

$$\lambda(E \cap X_n) = \int_{E \cap X_n} f_n \, d\mu = \int_E f_n \chi_{X_n} \, d\mu = \int_E f_n \, d\mu$$

for every $E \in \mathcal{X}$ (since $E \cap X_n \in \mathcal{X}_n$ and $f_n = f_n \chi_{X_n}$), for each $n \geq 1$. Now take an arbitrary set E in $\mathcal{X}$ and observe that $\{E \cap X_n\}$ is an increasing sequence of sets in $\mathcal{X}$ that covers E because $\{X_n\}$ is an increasing sequence of sets in $\mathcal{X}$ that covers X. Thus, according to Proposition 2.2(c),

$$\lambda(E) = \lambda\left(\bigcup_n (E \cap X_n)\right) = \lim_n (E \cap X_n) = \lim_n \int_E f_n \, d\mu.$$

The Monotone Convergence Theorem completes the existence proof. In fact, since $\{f_n\}$ is an increasing sequence of functions in $\mathcal{M}^+$, it follows by Theorem 3.4 that it converges pointwise to a function f' in $\mathcal{M}^+$ such that

$$\lambda(E) = \lim_n \int_E f_n \, d\mu = \lim_n \int f_n \chi_E \, d\mu = \int \lim_n f_n \chi_E \, d\mu = \int_E f' d\mu.$$

This function $f' \in \mathcal{M}^+$, being the pointwise limit of an increasing sequence of functions in $\mathcal{M}^+$, is possibly extended real-valued. However, it is real-valued μ-a.e. That is, we claim that the set $F_{+\infty} = \{x \in X : f'(x) = +\infty\}$ is such that $\mu(F_{+\infty}) = 0$. Indeed, recall that $\{X_n\}$ is an increasing sequence of sets that cover X, each f_n is null outside X_n, and $f_k = f_n$ μ-a.e. on X_n for all $k \geq n$. Then, since $f'(x) = \lim_n f_n(x)$ for every $x \in X$, it follows that $f_n = f' \chi_{X_n}$ μ-a.e., and so $\mu(F_{+\infty} \cap X_n) = \mu(\{x \in X_n : f_n(x) = +\infty\})$, for each n. But f_n lies in $L^1(X, \mathcal{X}, \mu)$ because $g_n = f_n \chi_{X_n}$ lies in $L^1(X_n, \mathcal{X}_n, \mu)$. Thus $\mu(F_{+\infty} \cap X_n) = 0$ for all n according to Problem 3.9(b). Since $F_{+\infty} = F_{+\infty} \cap X = F_{+\infty} \cap \bigcup_n X_n = \bigcup_n (F_{+\infty} \cap X_n)$, it follows that $\mu(F_{+\infty}) \leq \sum_n \mu(F_{+\infty} \cap X_n) = 0$ (see Problem 2.8(b)), and hence $\mu(F_{+\infty}) = 0$. Put $f = f' \chi_{X \setminus F_{+\infty}}$, a real-valued function in $\mathcal{M}^+$ such that (see Problem 3.8)

$$\lambda(E) = \int_{E \cap (X \setminus F_{+\infty})} f' d\mu + \int_{E \cap F_{+\infty}} f' d\mu = \int_E f' \chi_{X \setminus F_{+\infty}} d\mu = \int_E f \, d\mu$$

for every $E \in \mathcal{X}$. This $f \in \mathcal{M}^+$ is unique μ-a.e. by Problem 3.8(d).   □

The real-valued function $f \in \mathcal{M}(X, \mathcal{X})^+$ in the statement of the Radon–Nikodým Theorem is not asserted to be integrable. In fact, it is clear that f is μ-integrable (i.e., f lies in $\mathcal{L}(X, \mathcal{X}, \mu)$) if and only if λ is a finite measure. This f is called the *Radon–Nikodým derivative* of λ with respect to μ, and we shall sometimes write $f = \frac{d\lambda}{d\mu}$ (or $d\lambda = f \, d\mu$). Again (see Problem 3.11) no independent meaning is assigned to the symbols $d\lambda$ and $d\mu$. Thus, if λ and μ are σ-finite measures such that $\lambda \ll \mu$, then there is a unique (μ-a.e.) real-valued function $\frac{d\lambda}{d\mu}$ in $\mathcal{M}(X, \mathcal{X})^+$ such that, for every E in $\mathcal{X}$,

$$\lambda(E) = \int_E \frac{d\lambda}{d\mu} \, d\mu.$$

Remark: A rather important application of the Radon–Nikodým Theorem is the *Riesz Representation Theorem*, which reads as follows. *If $G: L^p(\mu) \to \mathbb{R}$ is a bounded linear functional on the Banach space $L^p(\mu) = L^p(X, \mathcal{X}, \mu)$, then there exists a unique $g \in L^q(\mu)$ such that $G(f) = \int fg \, d\mu$ for every $f \in L^p(\mu)$ and $\|G\| = \|g\|_q$ (where q is the Hölder conjugate of p; if $p = 1$ so that $q = \infty$, then μ is supposed to be σ-finite).* The reader may be aware that the Riesz Representation Theorem holds in every Hilbert space, and so it can be proved for $p = 2$ without using the Radon–Nikodým Theorem and, perhaps surprisingly, this can be used to prove the Radon–Nikodým Theorem itself. See the references in the Suggested Reading section.

7.3   The Lebesgue Decomposition Theorem

If a measure λ is absolutely continuous with respect to a measure μ, then sets of small μ-measures have small λ-measures (Proposition 7.7). At the opposite end we might think of measures λ and μ where sets of small μ-measure have large λ-measure. This is the concept introduced next.

Definition 7.9. Let $(X, \mathcal{X})$ be a measurable space. A measure λ on $\mathcal{X}$ is *singular* with respect to a measure μ on $\mathcal{X}$ (notation: $\lambda \perp \mu$) if there exists a measurable partition $\{A, B\}$ of X such that

$$\lambda(A) = \mu(B) = 0.$$

This means that λ and μ have *disjoint supports* (see Problem 2.13), which is also referred to by saying that λ is *concentrated* on a set of μ-measure zero. It is plain that $\perp$ is a symmetric relation on the collection of all measures on $\mathcal{X}$ (i.e., $\lambda \perp \mu$ if and only if $\mu \perp \lambda$). For this reason we also say that λ and μ are *mutually singular*, or simply *singular*, instead of λ is singular with respect to μ (or vice versa). It is worth noticing that this is sometimes rephrased as follows: $\lambda \perp \mu$ if there exists a partition $\{A, B\}$ of X such that $A \cap E$ and $B \cap E$ lie in $\mathcal{X}$ for every $E \in \mathcal{X}$ and

$$\lambda(A \cap E) = \mu(B \cap E) = 0.$$

It is clear that the preceding two expressions are equivalent. Indeed A and B are measurable (since $X \in \mathcal{X}$), $A \cap E \subseteq A$, and $B \cap E \subseteq B$. Here is another important consequence of the Radon–Nikodým Theorem.

Theorem 7.10. (Lebesgue Decomposition Theorem). *Let $(X, \mathcal{X})$ be a measurable space. If λ and μ are σ-finite measures on $\mathcal{X}$, then there exists a unique pair of measures λ_a and λ_s on $\mathcal{X}$ such that $\lambda_a \ll \mu$, $\lambda_s \perp \mu$, and*

$$\lambda = \lambda_a + \lambda_s.$$

Proof. Put $\nu = \mu + \lambda$. This is a σ-finite measure on $\mathcal{X}$ since each μ and λ is σ-finite (Problem 2.14). Note that both μ and λ are absolutely continuous with respect to ν (i.e., if $\nu(E) = 0$ for some $E \in \mathcal{X}$, then $\mu(E) = \lambda(E) = 0$, and so $\mu \ll \nu$ and $\lambda \ll \nu$). Then the Radon–Nikodým Theorem says that

$$\mu(E) = \int_E f \, d\nu \quad \text{and} \quad \lambda(E) = \int_E g \, d\nu$$

for every $E \in \mathcal{X}$, for some real-valued functions f and g in $\mathcal{M}(X, \mathcal{X})^+$. Consider the measurable partition $\{F_0, F_+\}$ of X, with $F_0 = \{x \in X \colon f(x) = 0\}$ and $F_+ = \{x \in X \colon f(x) > 0\}$, and set $\lambda_s = \lambda_{F_0}$ and $\lambda_a = \lambda_{F_+}$, where the measures λ_{F_0} and λ_{F_+} on $\mathcal{X}$ were defined in Problem 2.11. That is,

$$\lambda_s(E) = \lambda_{F_0}(E) = \lambda(E \cap F_0) \quad \text{and} \quad \lambda_a(E) = \lambda_{F_+}(E) = \lambda(E \cap F_+)$$

for each $E \in \mathcal{X}$. Since $\mu(F_0) = \int_{F_0} 0 \, d\nu = 0$ and $\lambda_s(F_+) = \lambda(\varnothing) = 0$, we get

$$\lambda_s \perp \mu.$$

Now, if $E \in \mathcal{X}$ is such that $\mu(E) = 0$, then $\int f\chi_E \, d\nu = \int_E f \, d\nu = 0$, and so $f\chi_E = 0$ ν-a.e. (Proposition 3.7(a)). That is, $f = 0$ ν-a.e. on E, and hence $\nu(E \cap F_+) = 0$. Since $\lambda \ll \nu$, it follows that $\lambda_a(E) = \lambda(E \cap F_+) = 0$. Thus,

$$\lambda_a \ll \mu.$$

Observe that $\lambda(E) = \lambda((E \cap F_0) \cup (E \cap F_+)) = \lambda(E \cap F_0) + \lambda(E \cap F_+) = \lambda_s(E) + \lambda_a(E)$ for every $E \in \mathcal{X}$, which means

$$\lambda = \lambda_s + \lambda_a.$$

Finally, to prove uniqueness, suppose $\lambda = \lambda_1 + \lambda_2$, where λ_1 and λ_2 are (σ-finite) measures on $\mathcal{X}$ such that $\lambda_1 \perp \mu$ and $\lambda_2 \ll \mu$. Consider the signed measures $\lambda_s - \lambda_1$ and $\lambda_a - \lambda_2$ on $\mathcal{X}$ so that $\lambda_s - \lambda_1 \perp \mu$ and $\lambda_a - \lambda_2 \ll \mu$ (cf. Problems 7.10 and 7.11). Since $\lambda_s + \lambda_a = \lambda_1 + \lambda_2$, it follows by Problem 7.12 that $\lambda_s - \lambda_1 = \lambda_a - \lambda_2 = 0$, and hence $\lambda_1 = \lambda_s$ and $\lambda_2 = \lambda_a$. $\square$

Remark: The foregoing signed measures are well defined if we allow extended real-valued signed measures and if we declare that $(\lambda_s - \lambda_1)(E) = 0$ if E in $\mathcal{X}$ is such that $\lambda_s(E) = \lambda_1(E) = +\infty$ and $(\lambda_a - \lambda_2)(E) = 0$ if E in $\mathcal{X}$ is such that $\lambda_a(E) = \lambda_2(E) = +\infty$. Also note that Problems 7.10, 7.11, and 7.12 are naturally extended to extended real-valued signed measures.

Theorem 7.10 decomposes any σ-finite measure λ into two parts, one absolute continuous and the other singular, both with respect to a σ-finite *reference measure* μ (e.g., such a reference measure is often taken to be the Lebesgue measure on the Borel algebra $\mathfrak{R}$). This can go further as follows.

Definition 7.11. Let $(X, \mathcal{X})$ be a measurable space. A measure λ on $\mathcal{X}$ is *continuous* with respect to a measure μ on $\mathcal{X}$ if, for $\{x\} \in \mathcal{X}$,

$$\mu(\{x\}) = 0 \quad \text{implies} \quad \lambda(\{x\}) = 0$$

(i.e., $\lambda(\{x\}) = 0$ for every *measurable* singleton $\{x\}$ for which $\mu(\{x\}) = 0$).

Definition 7.12. A measure λ on $\mathcal{X}$ is *discrete* with respect to measure μ on $\mathcal{X}$ if there is a measurable partition $\{A, B\}$ of X such that B is a countable set with all subsets measurable (i.e., $B = \{b_n\}_{n \in I}$ with each singleton $\{b_n\}$ lying in $\mathcal{X}$, where the index set I is either finite or $I = \mathbb{N}$) and

$$\lambda(A) = \mu(B) = 0$$

(i.e., λ is concentrated on a *countable* set of μ-measure zero).

Proposition 7.13. *Let $(X, \mathcal{X})$ be a measurable space. If λ and μ are measures on $\mathcal{X}$, where λ is σ-finite and measurable singletons of X have*

μ-measure zero, then there is a unique pair of measures λ_c and λ_d on $\mathcal{X}$ such that λ_c is continuous and λ_d is discrete, both with respect to μ, and

$$\lambda = \lambda_c + \lambda_d.$$

Proof. Since λ is σ-finite, there exists a sequence $\{E_n\}$ of sets in $\mathcal{X}$ that cover X such that each E_n has finite μ-measure. For each $k \geq 1$ put

$$B_k(n) = \{x \in E_n \colon \{x\} \in \mathcal{X} \text{ and } \lambda(\{x\}) \geq \tfrac{1}{k}\}.$$

If $B_k(n)$ is an infinite set, then it has a countably infinite subset, say $C_k(n) = \bigcup_m \{b_m\}$, consisting of distinct points b_m of $B_k(n)$. Since each singleton $\{b_m\}$ is $\mathcal{X}$-measurable, it follows that $C_k(n)$ also lies in $\mathcal{X}$, and hence $\lambda(C_k(n)) = \sum_m \lambda(\{b_m\}) = \infty$ because $\lambda(\{b_m\}) \geq \tfrac{1}{k}$ for all m. But this contradicts the fact that $\lambda(C_k(n)) \leq \lambda(E_n) < \infty$. Outcome: each $B_k(n)$ is a finite set. Therefore, since $X = \bigcup_n E_n$,

$$B_k = \bigcup_n B_k(n) = \{x \in X \colon \{x\} \in \mathcal{X} \text{ and } \lambda(\{x\}) \geq \tfrac{1}{k}\}$$

is a countable set, and so

$$B = \bigcup_k B_k = \{x \in X \colon \{x\} \in \mathcal{X} \text{ and } \lambda(\{x\}) \neq 0\}$$

is again a countable set (because a countable union of countable sets is countable). Note that B is measurable (i.e., $B \in \mathcal{X}$, since B is a countable union of measurable singletons) and consider the measurable partition $\{A, B\}$ of X with $A = X \backslash B$. Set $\lambda_c = \lambda_A$ and $\lambda_d = \lambda_B$, where the measures λ_A and λ_B on $\mathcal{X}$ were defined in Problem 2.11. That is,

$$\lambda_d(E) = \lambda_B(E) = \lambda(E \cap B) \quad \text{and} \quad \lambda_c(E) = \lambda_A(E) = \lambda(E \cap A)$$

for each $E \in \mathcal{X}$. Since $B = \{b_n\}_{n \in I}$ is a countable set consisting of measurable singletons, and since measurable singletons have μ-measure zero,

$$\lambda_d(A) = \lambda(A \cap B) = 0 \quad \text{and} \quad \mu(B) = \mu\left(\bigcup_n \{b_n\}\right) = \sum_n \mu(\{b_n\}) = 0,$$

and hence λ_d is discrete with respect to μ. Now, if $\{x\}$ is a measurable singleton of X, then either $\lambda(\{x\}) = 0$ or $\lambda(\{x\}) \neq 0$. In the former case, $\{x\} \subseteq A$ so that $\lambda_c(\{x\}) = \lambda(\{x\}) = 0$. In the latter case, $\{x\} \subseteq B$ so that $\lambda_c(\{x\}) = \lambda(\varnothing) = 0$. Thus $\lambda_c(\{x\}) = 0$ for all $\{x\} \in \mathcal{X}$, and so λ_c is continuous with respect to μ because $\mu(\{x\}) = 0$ for all $\{x\} \in \mathcal{X}$. Moreover,

$$\lambda(E) = \lambda((E \cap A) \cup (E \cap B)) = \lambda(E \cap A) + \lambda(E \cap B) = \lambda_c(E) + \lambda_d(E)$$

for every $E \in \mathcal{X}$, which means that

$$\lambda = \lambda_c + \lambda_d.$$

Finally, to prove uniqueness, suppose $\lambda = \lambda_1 + \lambda_2$, where λ_1 and λ_2 are measures on $\mathcal{X}$ such that λ_1 is continuous and λ_2 is discrete, both with respect to μ. Take a singleton $\{x\}$ in $\mathcal{X}$. Since $\mu(\{x\}) = 0$, it follows that $\lambda_1(\{x\}) = \lambda_c(\{x\}) = 0$, and so $\lambda_2(\{x\}) = \lambda(\{x\}) = \lambda_d(\{x\})$. Thus $\lambda_2 = \lambda_d$ by Problem 7.14. If $\lambda_1 \neq \lambda_c$, then there is a measurable set $E \subseteq A$ such that $\lambda_1(E) \neq \lambda_c(E)$. But $\lambda_2(E) = \lambda_d(E) = 0$ because $E \subseteq A$ and $\lambda_2 = \lambda_d$, and hence $\lambda(E) = \lambda_1(E) = \lambda_c(E)$, a contradiction. Therefore, $\lambda_1 = \lambda_c$. □

Corollary 7.14. *Let $(X, \mathcal{X})$ be a measurable space. Suppose μ is a σ-finite measure on $\mathcal{X}$ such that measurable singletons of X have μ-measure zero. Every σ-finite measure λ on $\mathcal{X}$ has a unique decomposition*

$$\lambda = \lambda_a + \lambda_{sc} + \lambda_{sd}$$

(called canonical decomposition), where λ_a, λ_{sc}, and λ_{sd} are measures on $\mathcal{X}$ such that λ_a is absolutely continuous, λ_{sc} is both singular and continuous (called singular-continuous), and λ_{sd} is both singular and discrete (called singular-discrete), all with respect to μ.

Proof. $\lambda = \lambda_a + \lambda_s$ by Theorem 7.10, where λ_a is absolutely continuous and λ_s is singular, with respect μ. Since λ_s is σ-finite (use the same countable covering of X that makes λ σ-finite), $\lambda_s = \lambda_{sc} + \lambda_{sd}$ by Proposition 7.13, where λ_{sc} is continuous and λ_{sd} is discrete, with respect to μ. Since $\lambda_s \perp \mu$, it follows that there exists a measurable partition $\{A, B\}$ of X such that $\lambda_s(A) = \lambda_{sc}(A) + \lambda_{sd}(A) = 0 = \mu(B)$, and hence $\lambda_{sc}(A) = \lambda_{sd}(A) = 0$, so λ_{sc} and λ_{sd} also are singular with respect to μ.   □

7.4   Problems

Problem 7.1. Let $\nu \colon \mathcal{X} \to \mathbb{R}$ be a signed measure on $\mathcal{X}$. If A and B are $\mathcal{X}$-measurable sets and $\{E_n\}$ is a sequence of sets in $\mathcal{X}$, then show that

   (a) $\nu(B \backslash A) = \nu(B) - \nu(A)$ if $A \subseteq B$,

   (b) $\nu\left(\bigcup_n E_n\right) = \lim_n \nu(E_n)$ if $\{E_n\}$ is increasing,

   (c) $\nu\left(\bigcap_n E_n\right) = \lim_n \nu(E_n)$ if $\{E_n\}$ is decreasing.

Hint: Proof of Proposition 2.2 and Definition 2.3.

Problem 7.2. (a) Every measurable subset of a positive set is positive. (b) A countable union of positive sets is again a positive set. Prove.

Hint: (b) Take a sequence $\{A_n\}$ of sets. Put $A'_{n+1} = A_{n+1} \backslash \left(\bigcup_{i=1}^n A_i\right)$ with $A'_1 = A_1$. This $\{A'_n\}$ is a sequence of disjoint sets such that $\bigcup_n A'_n = \bigcup_n A_n$

(i.e., $\{A'_n\}$ is a *disjointification* of $\{A_n\}$). Suppose each A_n is a positive set. Each A'_n is a measurable subset of A_n and so is a positive set by (a). Thus, $\nu(E \cap \bigcup_n A_n) = \nu(E \cap \bigcup_n A'_n) = \nu(\bigcup_n(E \cap A'_n)) = \sum_n \nu(E \cap A'_n) \geq 0$.

Problem 7.3. Let ν be a signed measure on a σ-algebra $\mathcal{X}$. If A and B lie in $\mathcal{X}$ and B is positive with respect to ν, then (Problems 7.1(a) and 7.2(a))

$$A \subseteq B \quad \text{implies} \quad 0 \leq \nu(A) \leq \nu(B).$$

Problem 7.4. (Signed-measure version of Problem 2.8). Let $\nu: \mathcal{X} \to \mathbb{R}$ be a signed measure and let $\{E_n\}$ be a sequence of measurable sets. Show that

(a) $\nu(\bigcup_n E_n) = \lim_n \nu(\bigcup_{i=1}^n E_i)$.

If each E_n is positive with respect to ν, then

(b) $\nu(\bigcup_n E_n) \leq \sum_n \nu(E_n)$.

Hints: (a) Problem 7.1(b). (b) Disjointification and Problem 7.3.

Problem 7.5. Let ν be a signed measure on $\mathcal{X}$. Take any set E in $\mathcal{X}$ and put $\mathcal{E} = \wp(E) \cap \mathcal{X}$, the σ-algebra of all measurable subsets of E. Show that

$$\nu^+(E) = \sup_{F \in \mathcal{E}} \nu(F) \quad \text{and} \quad \nu^-(E) = -\inf_{F \in \mathcal{E}} \nu(F).$$

Hint: Theorem 7.4: $\nu(F) = \nu^+(F) - \nu^-(F) \leq \nu^+(F) \leq \nu^+(E) = \nu(A^+ \cap E)$.

Problem 7.6. Let ν be a signed measure on $\mathcal{X}$. Show that $N \in \mathcal{X}$ is a null set with respect to ν if and only if $|\nu|(N) = 0$.

Hint: Definition of $|\nu|$ on the one hand; Problem 7.5 on the other hand.

Problem 7.7. Let $(X, \mathcal{X})$ be a measurable space. If ν, ν_1, and ν_2 are signed measures on $\mathcal{X}$ and α is any real number, then $\nu_1 + \nu_2$ and $\alpha\nu$ are again signed measures on $\mathcal{X}$, where, for each $E \in \mathcal{X}$,

$$(\alpha\nu)(E) = \alpha\nu(E) \quad \text{and} \quad (\nu_1 + \nu_2)(E) = \nu_1(E) + \nu_2(E).$$

Let $\mathcal{S} = \mathcal{S}(X, \mathcal{X}, \mathbb{R})$ denote the collection of all signed measures on $\mathcal{X}$. Since addition and scalar multiplication of signed measures are again signed measures, it follows that $\mathcal{S}$ is a (real) linear space (indeed, a linear subspace of the real linear space $\mathcal{X}^{\mathbb{R}}$ of all real-valued set functions on $\mathcal{X}$). First consider the total variation of signed measures in $\mathcal{S}$ and show that

(a) $|\alpha\nu| = |\alpha||\nu|$,

(b) $|\nu_1 + \nu_2| \leq |\nu_1| + |\nu_2|$.

Hint: Let $\{A_\nu^+, A_\nu^-\}$ be a Hahn decomposition of X with respect to ν. Verify that $\{A_{\alpha\nu}^+, A_{\alpha\nu}^-\}$ is a Hahn decomposition of X with respect to $\alpha\nu$, where $A_{\alpha\nu}^+ = A_\nu^+$ and $A_{\alpha\nu}^- = A_\nu^-$ if $\alpha \geq 0$ or $A_{\alpha\nu}^+ = A_\nu^-$ and $A_{\alpha\nu}^- = A_\nu^+$ if $\alpha \leq 0$. Then show that $(\alpha\nu)^+ = \alpha\nu^+$ and $(\alpha\nu)^- = \alpha\nu^-$ if $\alpha \geq 0$ or $(\alpha\nu)^+ = -\alpha\nu^-$ and $(\alpha\nu)^- = -\alpha\nu^+$ if $\alpha \leq 0$. Thus conclude the identity in (a): $|\alpha\nu| = (\alpha\nu)^+ + (\alpha\nu)^- = |\alpha|(\nu^+ + \nu^-) = |\alpha||\nu|$. To show the inequality in (b) note that $\nu_1 + \nu_2 = (\nu_1^+ + \nu_2^+) - (\nu_1^- + \nu_2^-)$, apply Theorem 7.4 again to verify that $(\nu_1 + \nu_2)^+ \leq \nu_1^+ + \nu_2^+$ and $(\nu_1 + \nu_2)^- \leq \nu_1^- + \nu_2^-$, and hence $|\nu_1 + \nu_2| = (\nu_1 + \nu_2)^+ + (\nu_1 + \nu_2)^- \leq (\nu_1^+ + \nu_1^-) + (\nu_2^+ + \nu_2^-) = |\nu_1| + |\nu_2|$.

Now consider the function $\|\ \|: \mathcal{S} \to \mathbb{R}$ defined by

$$\|\nu\| = |\nu|(X)$$

for every $\nu \in \mathcal{S}$. This is a norm on $\mathcal{S}$. That is, show that

(c) $(\mathcal{S}, \|\ \|)$ is a normed space.

Hint: Use (a) and (b) to verify axioms (iii) and (iv) of Definition 5.1.

Moreover, this normed space is complete. That is, show that

(d) $(\mathcal{S}, \|\ \|)$ is a Banach space.

Hint: Consider the following well-known result from elementary functional analysis. *A normed space is a Banach space if and only if every absolutely summable sequence is summable* (cf. Suggested Readings for Chapter 5). Note that, if $\{\nu_n\}$ is a sequence of signed measures in $\mathcal{S}$, then

$$\max\{\nu_n^+(E), \nu_n^-(E)\} \leq \nu_n^+(E) + \nu_n^-(E) \leq \nu_n^+(X) + \nu_n^-(X) = |\nu_n|(X),$$

where $0 \leq \min\{\nu_n^+(E), \nu_n^-(E)\}$, so that $\nu_n(E) = \nu_n^+(E) - \nu_n^-(E)$ makes a summable sequence for every E in $\mathcal{X}$ (i.e., $\{\nu_n\}$ is summable in $\mathcal{S}$) whenever $\{\|\nu_n\|\}$ is a summable in $\mathbb{R}$ (i.e., whenever $\{\nu_n\}$ is absolutely summable).

Problem 7.8. Let $(X, \mathcal{X})$ be a measurable space. Verify that absolute continuity $\ll$ is a reflexive and transitive but not symmetric relation on the collection of all measures on $\mathcal{X}$. That is, if λ, μ, and ν are measures on $\mathcal{X}$, then show that $\mu \ll \mu$ (reflexivity), and that $\lambda \ll \mu$ and $\mu \ll \nu$ imply $\lambda \ll \nu$ (transitivity), but $\lambda \ll \mu$ does not imply $\mu \ll \lambda$. If $\lambda \ll \mu$ and $\mu \ll \lambda$, then λ and μ are called *equivalent measures*. Notations: $\lambda \equiv \mu$ or $\lambda \sim \mu$.

Problem 7.9. Take a measure space $(X, \mathcal{X})$ and let λ, μ, and ν be σ-finite measures on $\mathcal{X}$. Prove the following propositions.

(a) If $\lambda \ll \mu$ and $g \in \mathcal{M}(X, \mathcal{X})^+$, then $\int_E g\, d\lambda = \int_E g \frac{d\lambda}{d\mu}\, d\mu$ for every $E \in \mathcal{X}$.

Hint: Theorem 7.8 and Problem 3.11 (recall: $g\chi_E \in \mathcal{M}(X, \mathcal{X})^+$).

(b) If $\lambda \ll \nu$ and $\mu \ll \nu$, then $\frac{d(\lambda+\mu)}{d\nu} = \frac{d\lambda}{d\nu} + \frac{d\mu}{d\nu}$ ν-almost everywhere.

  Hint: Theorem 7.8, Proposition 3.5(b), and Problem 3.8(d).

(c) If $\lambda \ll \mu \ll \nu$, then $\frac{d\lambda}{d\nu} = \frac{d\lambda}{d\mu}\frac{d\mu}{d\nu}$ ν-almost everywhere.

  Hint: Recall that $\lambda \ll \nu$. Apply Theorem 7.8 for each relation $\ll$ followed by Problem 3.11 as in part (a). Then use Problem 3.8(d).

(d) If $\lambda \ll \mu$ and $\mu \ll \lambda$, then $\frac{d\lambda}{d\mu} = (\frac{d\mu}{d\lambda})^{-1}$ almost everywhere.

  Note: $\lambda \ll \mu \ll \lambda$ means $\lambda \equiv \mu$ (i.e., λ and μ are equivalent measures) so that μ-almost everywhere is equivalent to λ-almost everywhere.

  Hint: $\frac{d\lambda}{d\lambda}$ is the identity. Use part (a) with $\nu = \lambda$. Swap λ and μ.

Problem 7.10. A signed measure ν on a σ-algebra $\mathcal{X}$ is *absolutely continuous* with respect to a signed measure μ on $\mathcal{X}$ if, for $E \in \mathcal{X}$,

$$|\mu|(E) = 0 \quad \text{implies} \quad \nu(E) = 0$$

(i.e., $\nu(E) = 0$ for every $E \in \mathcal{X}$ such that $|\mu|(E) = 0$). Notation: $\lambda \ll \mu$. Show that the following assertions are pairwise equivalent.

(a) $\nu \ll \mu$.

(b) $\nu^+ \ll \mu$ and $\nu^- \ll \mu$.

(c) $|\nu| \ll |\mu|$.

Hint: Let $\{A^+, A^-\}$ be a Hahn decomposition of X with respect to ν. Verify that $|\mu|(E) = 0$ implies $|\mu|(A^+ \cap E) = |\mu|(A^- \cap E) = 0$ and, if (a) holds, this implies $\nu^+(E) = \nu^-(E) = 0$. Thus conclude that (a) implies (b). That (b) implies (a) follows from the fact that $\nu = \nu^+ - \nu^-$ (Theorem 7.4). Similarly, verify that (b) and (c) are equivalent because $|\nu| = \nu^+ + \nu^-$.

Now consider a third signed measure λ on $\mathcal{X}$ and show that

(d) $\lambda \ll \mu$ and $\nu \ll \mu$ imply $(\lambda + \nu) \ll \mu$.

Hint: $|\lambda + \nu| \le |\lambda| + |\nu|$ according to Problem 7.7(b).

Problem 7.11. A signed measure ν on a σ-algebra $\mathcal{X}$ is *singular* with respect to a signed measure μ on $\mathcal{X}$ (or μ and ν are *mutually singular*, or simply *singular*) — notation: $\nu \perp \mu$ — if their total variations $|\nu|$ and $|\mu|$ are singular measures on $\mathcal{X}$ (according to Definition 7.9). That is,

$$\nu \perp \mu \quad \text{if and only if} \quad |\nu| \perp |\mu|.$$

Since $|\nu| = \nu^+ + \nu^-$, verify that

(a) $\qquad\qquad \nu \perp \mu$ implies $\nu^+ \perp \mu$ and $\nu^- \perp \mu$.

Now consider a third signed measure λ on $\mathcal{X}$ and show that

(b) $\qquad\qquad \lambda \perp \mu$ and $\nu \perp \mu$ imply $(\lambda + \nu) \perp \mu$.

Hint: If $|\lambda|(A) = |\mu|(B) = 0$ and $|\nu|(C) = |\mu|(D)$, where $\{A, B\}$ and $\{C, D\}$ are measurable partitions of X, then $\{E, F\}$ forms another measurable partition of X, with $E = (A \cap C)$ and $F = (B \cap C) \cup (A \cap D) \cup (B \cap D)$, such that $|\lambda|(E) = |\nu|(E) = |\mu|(F) = 0$. Recall from Problem 7.7(b) that $|\lambda + \nu| \leq |\lambda| + |\nu|$, and so $|\lambda + \nu|(E) = \mu(F) = 0$.

Problem 7.12. Take a measurable space $(X, \mathcal{X})$. If λ and μ are measures (or signed measures) on $\mathcal{X}$ such that $\lambda \ll \mu$ and $\lambda \perp \mu$, then $\lambda = 0$.

Problem 7.13. Take a measurable space $(X, \mathcal{X})$, let λ and μ be measures on $\mathcal{X}$, and verify the following propositions.

(a) If $\lambda \ll \mu$, then λ is continuous with respect μ.

(b) If λ is discrete with respect to μ, then $\lambda \perp \mu$.

(c) If λ is continuous and discrete with respect to μ, then $\lambda = 0$.

Note that the term *singular-discrete* used in Corollary 7.14, although of common usage, is, in fact, a pleonasm according to (b).

Hint: Definitions 7.6, 7.9, and 7.11. (c) If λ is discrete and continuous with respect to μ, then $\lambda(X) = \lambda(A) + \lambda(B) = \lambda(B) = \sum_n \lambda(b_n) = 0$, since each $\{b_n\}$ is measurable and $\mu(\{b_n\}) \leq \mu(B) = 0$, so $\lambda(\{b_n\}) = 0$.

Problem 7.14. Let λ and μ be measures on $\mathcal{X}$. Set $\mathcal{E} = \wp(E) \cap \mathcal{X}$ for each $E \in \mathcal{X}$: all measurable subsets of E. If λ is discrete with respect to μ, then

$$\lambda(E) = \sum_{\{x\} \in \mathcal{E}} \lambda(\{x\}) \quad \text{for every} \quad E \in \mathcal{X}$$

and $\mu(\{x\}) = 0$ whenever $\{x\}$ in $\mathcal{X}$ is such that $\lambda(\{x\}) \neq 0$.

Hint: $\lambda(E) = \lambda(E \cap B)$, where B is a countable set of measurable singletons.

Suggested Reading

Bartle [3], Berberian [6], Halmos [13], Kelley–Srinivasan [15], Royden [22], Rudin [23], Shilov–Gurevich [24]. See also [18, Section 6.8]. For construction of singular-continuous measures in terms of the Cantor function see [20].

8
Extension

8.1 Outer Measure

We have seen in previous chapters some simple examples of measures. In particular, we have considered (but not properly constructed) the Lebesgue measure λ on the Borel algebra $\Re$, which is the σ-algebra of subsets of the real line $\mathbb{R}$ generated by the collection of all open intervals. In Example 2C we promised to prove later the existence and uniqueness of the Lebesgue measure $\lambda \colon \Re \to \overline{\mathbb{R}}$. We fulfill that promise in this chapter according to the following program. First we introduce the concept of a measure on an algebra (rather than on a σ-algebra) of subsets of set X. Then we consider the notion of an outer measure generated by that measure on an algebra, which is a set function on the power set $\wp(X)$, that is, on the collection of all subsets of X. Finally, we show that this outer measure induces a σ-algebra of subsets of X on which we can define a *bona fide* measure.

Recall from Definition 1.1 that the difference between an algebra and a σ-algebra of subsets of a set X is that in an algebra $\mathcal{A}$ any *finite union* of sets in $\mathcal{A}$ is required to remain in $\mathcal{A}$, while in a σ-algebra $\mathcal{X}$ it is imposed, in addition, that any *countable union* of sets in $\mathcal{X}$ must remain in $\mathcal{X}$. We shall now define the notion of a measure μ on an algebra $\mathcal{A}$. Since a countable union of sets in $\mathcal{A}$ is not necessarily in $\mathcal{A}$, countable additivity for μ will be restricted to countable families of sets in $\mathcal{A}$ whose union still lies in $\mathcal{A}$.

Definition 8.1. Let $\mathcal{A}$ be an algebra of subsets of a set X. An extended real-valued set function μ on $\mathcal{A}$,

115

$$\mu\colon \mathcal{A} \to \overline{\mathbb{R}},$$

is a *measure* (on the algebra $\mathcal{A}$) if it satisfies the following conditions.

(a) $\mu(\varnothing) = 0$.

(b) $\mu(E) \geq 0$ for every $E \in \mathcal{A}$.

(c) $\mu\left(\bigcup_n E_n\right) = \sum_n \mu(E_n)$

for every countable family $\{E_n\}$ of pairwise disjoint sets in $\mathcal{A}$ for which $\bigcup_n E_n$ lies in $\mathcal{A}$.

As commented before (following Definition 2.1), if the countable set $\{\mu(E_n)\}$ of nonnegative (extended) real numbers in (c) is infinite (countably infinite), then the (infinite) series $\sum_n \mu(E_n)$ either converges unconditionally to a real number (i.e., the real value of the sum does not depend on the order of the summands) or diverges to infinity. Properties of measures on a σ-algebra are naturally transferred to measures on an algebra up to the assumption $\bigcup_n E_n \in \mathcal{A}$ in axiom (c), which is not necessary for a measure on a σ-algebra. Let $\mathcal{A}$ be any algebra of subsets of a set X. A measure μ on $\mathcal{A}$ generates a set function μ^* on the power set $\mathcal{P}(X)$ as follows.

Definition 8.2. Let $\mu\colon \mathcal{A} \to \overline{\mathbb{R}}$ be a measure on an algebra $\mathcal{A}$ of subsets of a set X. For each subset S of X consider the collection $\mathcal{C}_S$ of all countable families $\{E_n\}$ of sets in $\mathcal{A}$ that cover S,

$$\mathcal{C}_S = \{\{E_n\}\colon E_n \in \mathcal{A} \text{ and } S \subseteq \textstyle\bigcup_n E_n\}.$$

The extended real-valued set function μ^* on $\mathcal{P}(X)$,

$$\mu^*\colon \mathcal{P}(X) \to \overline{\mathbb{R}},$$

given for each $S \in \mathcal{P}(X)$ by

$$\mu^*(S) = \inf_{\{E_n\} \in \mathcal{C}_S} \sum_n \mu(E_n),$$

is the *outer measure* generated by the measure μ on $\mathcal{A}$.

The term *outer measure*, although usual, is inappropriate once μ^* may be far from being a measure itself. For instance, it is not necessarily additive: it may happen that $\mu^*(A \cup B) \neq \mu^*(A) + \mu^*(B)$ for some disjoint sets A and B in $\mathcal{P}(X)$. (We shall comment on this later: Problems 8.16 and 8.22.) The lack of additivity is attenuated by *subadditivity*. Actually, outer measures are *countably subadditive* according to Proposition 8.3(e) below. However, μ^* has the advantage of being defined for every subset of X and inherits some properties of a measure, such as $\mu^*(\varnothing) = 0$ and $\mu^*(S) \geq 0$ for every subset S of X and, what is more important, $\mu^*(E) = \mu(E)$ for every $E \in \mathcal{A}$.

Proposition 8.3. *If μ^* is the outer measure generated by a measure μ on an algebra $\mathcal{A}$ of subsets of a set X, then*

(a) $\mu^*(\varnothing) = 0$,

(b) $\mu^*(S) \geq 0$ *for every* $S \in \mathcal{P}(X)$,

(c) $\mu^*(S_1) \leq \mu^*(S_2)$ *whenever* $S_1 \subseteq S_2 \subseteq X$,

(d) $\mu^*(E) = \mu(E)$ *for every* $E \in \mathcal{A}$,

(e) $\mu^*\left(\bigcup_n E_n\right) \leq \sum_n \mu^*(E_n)$

for every countable family $\{E_n\}$ *of subsets of* X.

Proof. Properties (a), (b), and (c) are trivially verified by the definition of outer measure. To prove property (d) take an arbitrary $E \in \mathcal{A}$. Consider the sequence $\{E'_n\}$ of subsets of $\mathcal{A}$ such that $E'_1 = E$ and $E'_n = \varnothing$ for all $n \neq 1$. Since $\{E'_n\} \in \mathcal{C}_E$, it follows that

$$\mu^*(E) = \inf_{\{E_n\} \in \mathcal{C}_E} \sum_n \mu(E_n) \leq \sum_n \mu(E'_n) = \mu(E).$$

On the other hand, if $\{E_n\}$ is a countable family in $\mathcal{C}_E$, then $\{E \cap E_n\}$ is again a countable family in $\mathcal{C}_E$ such that $E = \bigcup_n (E \cup E_n)$, and therefore (cf. Problem 2.8(b) and Proposition 2.2(b))

$$\mu(E) \leq \sum_n \mu(E \cup E_n) \leq \sum_n \mu(E_n).$$

Since this inequality holds for all $\{E_n\} \in \mathcal{C}_E$, it follows that

$$\mu(E) \leq \inf_{\{E_n\} \in \mathcal{C}_E} \sum_n \mu(E_n) = \mu^*(E),$$

which completes the proof of (d). Now take an arbitrary $\varepsilon > 0$. Let $\{E_n\}$ be any countable family of subsets of X and recall that, for each E_n,

$$\mu^*(E_n) = \inf_{\{F_k\} \in \mathcal{C}_{E_n}} \sum_k \mu(F_k).$$

Thus for each $n \geq 1$ there exists a countable family $\{F'_{n,k}\}$ in $\mathcal{C}_{E_n}$ for which

$$\sum_k \mu(F'_{n,k}) \leq \mu^*(E_n) + \tfrac{\varepsilon}{n^2}.$$

Observe that $\{F'_{n,k}\} = \bigcup_n \bigcup_k F'_{n,k}$ is a countable family of sets in $\mathcal{A}$ covering $\bigcup_n E_n$, which means that $\{F'_{n,k}\}$ lies in $\mathcal{C}_{\bigcup_n E_n}$. Since

$$\mu^*\left(\bigcup_n E_n\right) = \inf_{\{F_{n,k}\} \in \mathcal{C}_{\bigcup_n E_n}} \sum_{n,k} \mu(F_{n,k}),$$

it follows that

$$\mu^*\left(\bigcup_n E_n\right) \le \sum_{n,k} \mu(F'_{n,k}) = \sum_n \sum_k \mu(F'_{n,k}) \le \sum_n \mu^*(E_n) + \varepsilon \sum_n \frac{1}{n^2},$$

and hence, because this holds for every $\varepsilon > 0$ and $\sum_n \frac{1}{n^2} < \infty$,

$$\mu^*\left(\bigcup_n E_n\right) \le \inf_{\varepsilon > 0}\left(\sum_n \mu^*(E_n) + \varepsilon \sum_n \frac{1}{n^2}\right) = \sum_n \mu^*(E_n),$$

which proves (e). $\square$

8.2   Carathéodory and Hahn Extensions

Let μ^* be the outer measure generated by a measure μ on an algebra $\mathcal{A}$ of subsets of a set X. We say that a set $E \in \wp(X)$ is μ^*-*measurable* (or satisfies the *Carathéodory condition*) if

$$\mu^*(S) = \mu^*(S \cap E) + \mu^*(S\backslash E)$$

for every $S \in \wp(X)$. This means that μ^* behaves additively on E. Set

$$\mathcal{A}^* = \{E \in \wp(X)\colon E \text{ is } \mu^*\text{-measurable}\},$$

the collection of all μ^*-measurable subsets of X. The next result is crucial. It says that this $\mathcal{A}^*$ is a σ-algebra such that $\mathcal{A} \subseteq \mathcal{A}^*$, and the restriction of the outer measure μ^* to $\mathcal{A}^*$ is a measure (on the σ-algebra $\mathcal{A}^*$) that extends the measure μ (on the algebra $\mathcal{A}$) over $\mathcal{A}^*$.

Theorem 8.4. (Carathéodory Extension Theorem). $\mathcal{A}^*$ *is a σ-algebra that includes the algebra $\mathcal{A}$, and the restriction of μ^* to $\mathcal{A}^*$ is a measure on $\mathcal{A}^*$.*

Proof. The empty set $\varnothing$ and the whole set X clearly lie in $\mathcal{A}^*$ (i.e., they are μ^*-measurable). Moreover, since $S \cap E = S\backslash(X\backslash E)$ for every pair of sets E and S in $\wp(X)$, it follows that the complement of every set in $\mathcal{A}^*$ is again in $\mathcal{A}^*$. Now take an arbitrary $S \in \wp(X)$. If E and F lie in $\mathcal{A}^*$, then

$$\mu^*(S) = \mu^*(S \cap F) + \mu^*(S\backslash F)$$

since $F \in \mathcal{A}^*$, and

$$\mu^*(S \cap F) = \mu^*(S \cap F \cap E) + \mu^*((S \cap F)\backslash E)$$

since $E \in \mathcal{A}^*$. But $[S\backslash(E \cap F)] \cap F = (S \cap F)\backslash(E \cap F) = (S \cap F)\backslash E$ and $[S\backslash(E \cap F)]\backslash F = (S\backslash F)\backslash(E \cap F) = S\backslash F$. Thus, since $F \in \mathcal{A}^*$,

$$\mu^*(S\backslash(E\cap F)) = \mu^*([S\backslash(E\cap F)]\cap F) + \mu^*([S\backslash(E\cap F)]\backslash F)$$
$$= \mu^*((S\cap F)\backslash E) + \mu^*(S\backslash F).$$

The above three identities ensure that

$$\mu^*(S) = \mu^*(S\cap F\cap E) + \mu^*((S\cap F)\backslash E) + \mu^*(S\backslash F)$$
$$= \mu^*(S\cap E\cap F) + \mu^*(S\backslash(E\cap F)),$$

and hence $E\cap F$ lies in $\mathcal{A}^*$. Since intersection of sets in $\mathcal{A}^*$ and complements of sets in $\mathcal{A}^*$ are both again in $\mathcal{A}^*$, it follows that union of sets in $\mathcal{A}^*$ also lie in $\mathcal{A}^*$ (for $E\cup F = X\backslash[(X\backslash E)\cap(X\backslash F)]$ — De Morgan Laws). Thus a trivial induction ensures that any finite union of sets in $\mathcal{A}^*$ remains in $\mathcal{A}^*$, and so $\mathcal{A}^*$ is an algebra. That is,

(a) $\bigcup_{i=1}^{n} F_n$ lies in $\mathcal{A}^*$ for every finite family $\{F_n\}_{i=1}^{n}$ of sets in $\mathcal{A}^*$.

Moreover, if S and F are sets in $\wp(X)$, E is a set $\mathcal{A}^*$, and E and F are disjoint (so that $S\cap(E\cup F)\backslash E = S\cap F$), then

$$\mu^*(S\cap(E\cup F)) = \mu^*(S\cap(E\cup F)\cap E) + \mu^*(S\cap(E\cup F)\backslash E)$$
$$= \mu^*(S\cap E) + \mu^*(S\cap F),$$

and μ^* acts additively on the intersection of any set in $\wp(X)$ with every pair of disjoint sets in $\mathcal{A}^*$. Thus another trivial induction ensures that

(b) $\mu^*\big(S\cap\big(\bigcup_{i=1}^{n} E_i\big)\big) = \sum_{i=1}^{n}\mu^*(S\cap E_i)$ for every finite family $\{E_i\}_{i=1}^{n}$ of pairwise disjoint sets in $\mathcal{A}^*$ and every set S in $\wp(X)$.

In particular, with $S = X$, this shows that μ^* is *finitely additive* on $\mathcal{A}^*$:

(c) $\mu^*\big(\bigcup_{i=1}^{n} E_i\big) = \sum_{i=1}^{n}\mu^*(E_i)$ for every finite family $\{E_i\}_{i=1}^{n}$ of pairwise disjoint sets in $\mathcal{A}^*$.

We now extend these results on finite families to countably infinite families, thus showing that $\mathcal{A}^*$ is a σ-algebra and μ^* is countably additive on $\mathcal{A}^*$, and so the restriction of μ^* to $\mathcal{A}^*$ is in fact a measure (according to Definition 2.1). Take any (infinite) sequence $\{F_n\}$ of sets in $\mathcal{A}^*$, and consider the disjointification $\{E_n\}$ of $\{F_n\}$ given by $E_1 = F_1$ and $E_{n+1} = F_{n+1}\backslash\big(\bigcup_{i=1}^{n} F_i\big)$ (cf. Hints to Problems 2.3 and 7.2). Observe that $\{E_n\}$ is a sequence of pairwise disjoint sets in $\mathcal{A}^*$ (each E_n lies in $\mathcal{A}^*$ because finite union and difference of sets in an algebra remain there). For each integer $n\geq 1$ put $G_n = \bigcup_{i=1}^{n} E_i = \bigcup_{i=1}^{n} F_i$, which lies in $\mathcal{A}^*$ (finite union of sets in $\mathcal{A}^*$), and set $G = \bigcup_{n=1}^{\infty} E_n = \bigcup_{n=1}^{\infty} F_n$ in $\wp(X)$. Take an arbitrary S in $\wp(X)$. First note that, by the countable subadditivity of Proposition 8.3(e),

$$\mu^*(S\cap G) = \mu^*\Big(S\cap\bigcup_{i=1}^{\infty} E_i\Big) = \mu^*\Big(\bigcup_{i=1}^{\infty}(S\cap E_i)\Big) \leq \sum_{i=1}^{\infty}\mu^*(S\cap E_i),$$

and so, since $S = (S \cap G) \cup (S \backslash G)$, subadditivity ensures again that

$$\mu^*(S) \leq \mu^*(S \cap G) + \mu^*(S \backslash G) \leq \sum_{i=1}^{\infty} \mu^*(S \cap E_i) + \mu^*(S \backslash G).$$

On the other hand, since each $G_n = \bigcup_{i=1}^{n} E_i$ is in $\mathcal{A}^*$ and $\{E_n\}$ is made up of pairwise disjoint sets in $\mathcal{A}^*$, it follows by (b) that

$$\mu^*(S) = \mu^*(S \cap G_n) + \mu^*(S \backslash G_n) = \sum_{i=1}^{n} \mu^*(S \cap E_i) + \mu^*(S \backslash G_n)$$

for all n. Since $G_n \subseteq G$, we get $S \backslash G \subseteq S \backslash G_n$, so $\mu^*(S \backslash G) \leq \mu^*(S \backslash G_n)$ by Proposition 8.3(c), and hence

$$\sum_{i=1}^{n} \mu^*(S \cap E_i) + \mu^*(S \backslash G) \leq \mu^*(S)$$

for all n, which implies that

$$\sum_{i=1}^{\infty} \mu^*(S \cap E_i) + \mu^*(S \backslash G) \leq \mu^*(S).$$

Therefore, for every $S \in \wp(X)$,

$$\mu^*(S) = \mu^*(S \cap G) + \mu^*(S \backslash G) = \sum_{i=1}^{\infty} \mu^*(S \cap E_i) + \mu^*(S \backslash G).$$

The first of the above two identities ensures that G lies in $\mathcal{A}^*$, that is,

(a') $\bigcup_n F_n$ lies in $\mathcal{A}^*$ for every countable family $\{F_n\}$ (not necessarily pairwise disjoint) of sets in $\mathcal{A}^*$,

which proves that the algebra $\mathcal{A}^*$ in fact is a σ-algebra. Moreover, taking $S = G$ (so that $\mu^*(S \backslash G) = 0$ and $S \cap E_i = E_i$ for every $i \geq 1$), we also get

$$\mu^*(G) = \sum_{i=1}^{\infty} \mu^*(E_i).$$

If $\{F_n\}$ is itself a sequence of pairwise disjoint sets, then $E_n = F_n$ for each n, and so the above identity ensures that μ^* is countably additive on $\mathcal{A}^*$:

(c') $\mu^*\left(\bigcup_n E_n\right) = \sum_n \mu^*(E_n)$ for every countably infinite family $\{E_n\}$ of pairwise disjoint sets in $\mathcal{A}^*$.

By (a') and (c'), the restriction of μ^* to $\mathcal{A}^*$ is a measure on the σ-algebra $\mathcal{A}^*$. Finally we show that $\mathcal{A} \subseteq \mathcal{A}^*$. Take an arbitrary set A in $\mathcal{A}$. Also take an arbitrary set S in $\wp(X)$ and any positive number ε. According to Definition

8.2, there exists a sequence $\{A_n\}$ of sets in $\mathcal{A}$ such that

$$S \subseteq \bigcup_n A_n \quad \text{and} \quad \sum_n \mu(A_n) \le \mu^*(S) + \varepsilon.$$

Observe that, for each n, $\{(A_n \cap A), (A_n \backslash A)\}$ is a partition of A_n consisting of sets in the algebra $\mathcal{A}$. Thus, according to Proposition 8.3(c,e,d),

$$\mu^*(S \cap A) \le \mu^*\left(\bigcup_n (A_n \cap A)\right) \le \sum_n \mu^*(A_n \cap A) = \sum_n \mu(A_n \cap A),$$

$$\mu^*(S \backslash A) \le \mu^*\left(\bigcup_n (A_n \backslash A)\right) \le \sum_n \mu^*(A_n \backslash A) = \sum_n \mu(A_n \backslash A),$$

and hence, using the additivity of Definition 8.1(c),

$$\mu^*(S \cap A) + \mu^*(S \backslash A) \le \sum_n \left(\mu(A_n \cap A) + \mu(A_n \backslash A)\right)$$

$$= \sum_n \mu\left((A_n \cap A) \cup (A_n \backslash A)\right) = \sum_n \mu(A_n) \le \mu^*(S) + \varepsilon.$$

Since $\varepsilon > 0$ is arbitrary, this implies that

$$\mu^*(S \cap A) + \mu^*(S \backslash A) \le \mu^*(S).$$

On the other hand, since $S = (S \cap A) \cup (S \backslash A)$,

$$\mu^*(S) \le \mu^*(S \cap A) + \mu^*(S \backslash A)$$

by the subadditivity of Proposition 8.3(e). Therefore,

$$\mu^*(S) = \mu^*(S \cap A) + \mu^*(S \backslash A)$$

for every $S \in \wp(X)$; that is, A lies in $\mathcal{A}^*$. Then $A \subseteq \mathcal{A}^*$. $\square$

Since $\mu^*(E) = \mu(E)$ for every set E in the algebra $\mathcal{A} \subseteq \mathcal{A}^* \subseteq \wp(X)$ and since the restriction of the outer measure $\mu^* : \wp(X) \to \overline{\mathbb{R}}$ to the σ-algebra $\mathcal{A}^*$ is a measure, it follows that this measure on the σ-algebra $\mathcal{A}^*$ actually extends the measure μ on the algebra $\mathcal{A}$ over A^*. Moreover, such a measure is complete in the sense of Section 2.3 or, equivalently, *the σ-algebra $\mathcal{A}^*$ is complete* with respect to it, which means that if $N \in \mathcal{A}^*$ and $\mu^*(N) = 0$, then every subset E of N lies in $\mathcal{A}^*$ (and, consequently, $\mu^*(E) = 0$ by Proposition 8.3(b,c)). As a matter of fact, more is true: *all sets with outer measure zero are μ^*-measurable* (and so $\mathcal{A}^*$ is complete).

Proposition 8.5. *Let μ^* be an outer measure on $\wp(X)$ and let $\mathcal{A}^*$ be the collection of all μ^*-measurable subsets of X.*

(a) *If $N \in \wp(X)$ is such that $\mu^*(N) = 0$, then $N \in \mathcal{A}^*$.*

(b) *If $E \subseteq N \in \wp(X)$ and $\mu^*(N) = 0$, then $E \in \mathcal{A}^*$ and $\mu(E) = 0$.*

Proof. The result in (b) is a straightforward consequence of (a) and Proposition 8.3(b,c). The result in (a) says that every set with outer measure zero is μ^*-measurable. Indeed, take an arbitrary S in $\mathcal{P}(X)$ and suppose $N \in \mathcal{P}(X)$ is such that $\mu^*(N) = 0$. Since $S \cup N = (S \cap N) \cup (S \backslash N)$, it follows by Proposition 8.3(c,e) that

$$\mu^*(S) \leq \mu^*(S \cup N) \leq \mu^*(S \cap N) + \mu^*(S \backslash N) \leq \mu^*(N) + \mu^*(S) = \mu^*(S),$$

and hence $\mu^*(S) = \mu^*(S \cap N) + \mu^*(S \backslash N)$, which means that $N \in \mathcal{A}^*$.   □

As in the case of measures on σ-algebras, a measure μ on an algebra $\mathcal{A}$ of subsets of a set X is *finite* if $\mu(X) < \infty$. Similarly, μ is *σ-finite* if X is covered by a countable family of sets in $\mathcal{A}$ of finite measure, that is, if there exists a sequence $\{A_n\}$ of sets in $\mathcal{A}$ such that $\mu(A_n) < \infty$ for every n and $X = \bigcup_n A_n$. Finiteness and σ-finiteness are naturally extended from a measure μ on an algebra $\mathcal{A}$ to the associated outer measure μ^* on the power set $\mathcal{P}(X)$ by Proposition 8.3(d) since $X \in \mathcal{A} \subseteq \mathcal{P}(X)$. The next result says that if μ is σ-finite, then its extension over the σ-algebra $\mathcal{A}^*$ is unique.

Theorem 8.6. (Hahn Extension Theorem). *If a measure μ on the algebra $\mathcal{A}$ is σ-finite, then its extension to a measure on the σ-algebra $\mathcal{A}^*$ is unique.*

Proof. Let μ^* be the outer measure generated by a measure μ on an algebra $\mathcal{A}$ of subsets of a set X. According to Theorem 8.5, the collection $\mathcal{A}^*$ of all μ^*-measurable subsets of X is a σ-algebra such that $\mathcal{A} \subseteq \mathcal{A}^*$, and the restriction of μ^* to $\mathcal{A}^*$ is a measure that extends μ over $\mathcal{A}^*$. Suppose ν is a measure on $\mathcal{A}^*$ that extends μ over $\mathcal{A}^*$. That is, suppose ν is a measure on $\mathcal{A}^*$ such that $\nu(E) = \mu(E)$ for every $E \in \mathcal{A}$. We split the proof into two parts. The theorem is proved assuming that μ is a finite measure in part (a). This will be extended to the case when μ is σ-finite in part (b).

(a) Suppose μ is a finite measure. Note that any extension of μ also is finite. Indeed, since X lies in $\mathcal{A} \subseteq \mathcal{A}^* \subseteq \mathcal{P}(X)$, it follows that $\mu^*(X) = \nu(X) = \mu(X) < \infty$. Take an arbitrary set E in $\mathcal{A}^*$. Let $\{A_n\}$ be any sequence in $\mathcal{C}_E$. That is, let $\{A_n\}$ be any sequence of sets in $\mathcal{A}$ such that $E \subseteq \bigcup_n A_n$. (Note that these sequences exist; for instance, let $A_1 = X$.) Thus, according to Proposition 2.2(a) and Problem 2.8(b),

$$\nu(E) \leq \nu\left(\bigcup_n A_n\right) \leq \sum_n \nu(A_n) = \sum_n \mu(A_n).$$

Since this holds for every $\{A_n\} \in \mathcal{C}_E$, it follows by Definition 8.2 that

$$\nu(E) \leq \inf_{\{E_n\} \in \mathcal{C}_E} \sum_n \mu(E_n) = \mu^*(E).$$

Recalling that μ^* (restricted to $\mathcal{A}^*$) and ν are measures, thus additive by Definition 2.1(c), that $\mu^*(X) = \nu(X)$, and that these measures are finite, we get by the preceding inequality that

$$\mu^*(E) = \mu^*(X) - \mu^*(X\backslash E) = \nu(X) - \mu^*(X\backslash E) \leq \nu(X) - \nu(X\backslash E) = \nu(E).$$

Thus $\nu(E) = \mu^*(E)$ for all $E \in \mathcal{A}^*$, which proves uniqueness if μ is finite.

(b) Now suppose μ is σ-finite. Again, note that any extension of μ also is σ-finite. Indeed, if μ is σ-finite, then there exists a sequence $\{A_n\}$ of sets in $\mathcal{A} \subseteq \mathcal{A}^* \subseteq \mathcal{P}(X)$ for which $\mu^*(A_n) = \nu(A_n) = \mu(A_n) < \infty$ for each $n \geq 1$ and $X = \bigcup_n A_n$, and hence μ^* and ν also are σ-finite. Put $A'_n = \bigcup_{i=1}^n A_n$ so that $\{A'_n\}$ is an *increasing* sequence of sets of finite measure that cover X. Take an arbitrary E in $\mathcal{A}^*$. Since $\{E \cap A'_n\}$ is a sequence of sets in $\mathcal{A}^*$ (intersection and finite union of sets in $\mathcal{A}^*$ remain in $\mathcal{A}^*$) of finite measure (because $\mu^*(E \cap A'_n) \leq \mu^*(A'_n) < \infty$ and $\nu(E \cap A'_n) \leq \nu(A'_n) < \infty$), it follows from part (a) that, for each $n \geq 1$,

$$\mu^*(E \cap A'_n) = \nu(E \cap A'_n).$$

Therefore, since $\{E \cap A'_n\}$ is an increasing sequence of sets in $\mathcal{A}^*$ such that $E = \bigcup_n (E \cap A'_n)$, and since both ν and the restriction $\mu^*|_{\mathcal{A}^*}$ of μ^* to the σ-algebra $\mathcal{A}^*$ are measures on $\mathcal{A}^*$, we get by Proposition 2.2(c) that

$$\mu^*(E) = \lim_n \mu^*(E \cap A'_n) = \lim_n \nu(E \cap A'_n) = \nu(E).$$

Thus $\nu(E) = \mu^*(E)$ for all $E \in \mathcal{A}^*$, proving uniqueness if μ is σ-finite. $\quad\square$

Let the restriction $\mu^*|_{\mathcal{A}^*} : \mathcal{A}^* \to \overline{\mathbb{R}}$ of the outer measure $\mu^* : \mathcal{P}(X) \to \overline{\mathbb{R}}$ to $\mathcal{A}^*$ be denoted again by μ^*. The foregoing results are summarized as follows. *If μ is a σ-finite measure defined on an algebra $\mathcal{A}$, then there exist a σ-algebra $\mathcal{A}^*$ including $\mathcal{A}$ and a unique extension of μ to a measure on $\mathcal{A}^*$.* This measure on $\mathcal{A}^*$ is σ-finite and, by uniqueness, coincides with μ^* (i.e., there exists a unique measure μ^* on the σ-algebra $\mathcal{A}^*$ such that $\mu^*(E) = \mu(E)$ for every $E \in \mathcal{A}$). Moreover, *the measure space $(X, \mathcal{A}^*, \mu^*)$ is complete* (which means that the σ-algebra $\mathcal{A}^*$ is complete with respect to the measure μ^* or, equivalently, the measure μ^* is complete on the σ-algebra $\mathcal{A}^*$).

8.3   Lebesgue Measure

Consider the following four classes of (left-open) intervals of the real line.

Class $\mathcal{C}_1$: $\{(\alpha, \beta] \subseteq \mathbb{R} : \alpha, \beta \in \mathbb{R}, \alpha \leq \beta\}$.

Class C_2: $\{(-\infty, \beta] \subseteq \mathbb{R}: \beta \in \mathbb{R}\}$.

Class C_3: $\{(\alpha, +\infty) \subseteq \mathbb{R}: \alpha \in \mathbb{R}\}$.

Class C_4: $\{(-\infty, +\infty)\}$.

Let $\mathcal{I}$ be the family of all intervals of the real line that belong to one of the four classes. Note that the empty set $\varnothing$ is of class C_1 (for the case of $\alpha = \beta$) and that the only interval of class C_4 is $\mathbb{R}$ itself. It is clear that the intersection of any two sets in $\mathcal{I}$ is again a set in $\mathcal{I}$ and that the complement of any set in $\mathcal{I}$ is a finite union of disjoint sets in $\mathcal{I}$. This means that $\mathcal{I}$ is a *semialgebra*. But finite union of sets in $\mathcal{I}$ is not necessarily a set in $\mathcal{I}$, so $\mathcal{I}$ is not an algebra. However, the collection $\Im$ of all finite unions of sets in $\mathcal{I}$ is an algebra (cf. Problem 8.3).

Definition 8.7. On the collection $\Im$ of all *finite* unions of intervals from $\mathcal{I}$,

$$\Im = \{F \subseteq \mathbb{R}: F \text{ is a finite union of intervals from } \mathcal{I}\},$$

define an extended real-valued set function $\ell: \Im \to \overline{\mathbb{R}}$ such that

(a) $\ell((\alpha, \beta]) = \beta - \alpha$ for any $\alpha, \beta \in \mathbb{R}$ such that $\alpha \leq \beta$,

(b) $\ell((-\infty, \beta]) = \ell((\alpha, +\infty)) = \ell((-\infty, +\infty)) = \infty$,

(c) $\ell(\bigcup_i I_i) = \sum_i \ell(I_i)$

for every *finite* family $\{I_i\}$ of pairwise disjoint intervals in $\mathcal{I}$,

which is referred to as the *length function* on $\Im$.

Indeed, properties (a), (b), and (c) are enough to make the length function $\ell: \Im \to \overline{\mathbb{R}}$ well defined on the whole collection $\Im$, which in fact is an algebra, and ℓ is a measure on it. To prove this we proceed as follows. First we check that ℓ is well defined at every set in $\Im$ (Problem 8.2). Next we verify that $\Im$ is an algebra (Problem 8.3), and so we conclude that the set function ℓ is increasing (Problem 8.4). This is applied to prove an auxiliary result in Proposition 8.8, which in turn is used to prove in Lemma 8.9 that the length function ℓ is countably additive, and hence it is a measure on $\Im$.

Proposition 8.8. *If* $\{(\alpha_k, \beta_k]\}$ *is a countable family of disjoint intervals of class* C_1 *and* $(a, b]$ *also is an interval of class* C_1, *then*

(a) $(a, b] = \bigcup_k (\alpha_k, \beta_k]$ implies $\ell((a, b]) = \sum_k \ell((\alpha_k, \beta_k])$.

Moreover, if $\{(a_i, b_i]\}$ *is a finite set of disjoint intervals of class* C_1, *then*

(b) $\bigcup_i (a_i, b_i] = \bigcup_k (\alpha_k, \beta_k]$ implies $\ell(\bigcup_i (a_i, b_i]) = \sum_k \ell((\alpha_k, \beta_k])$.

Proof. If the countable family $\{(\alpha_k, \beta_k]\}$ of disjoint intervals is finite, then the results in (a) and (b) are trivially verified by Definition 8.7(c). Thus suppose it is a countably infinite family. To avoid trivialities, let all intervals be *nonempty*, which means that $\alpha_k < \beta_k$ for every k.

(a) Suppose $(a, b] = \bigcup_k (\alpha_k, \beta_k]$. Take any finite subfamily $\{(\alpha_i, \beta_i]\}$ of the infinite family $\{(\alpha_k, \beta_k]\}$. Since $\{(\alpha_i, \beta_i]\}$ has a finite number of subintervals of $(a, b]$, it follows that $a \leq \min\{\alpha_i\}$ and $\max\{\beta_i\} \leq b$, and hence

$$\sum_i \ell((\alpha_i, \beta_i]) = \sum_i (\beta_i - \alpha_i) \leq b - a = \ell((a, b]).$$

This holds for all finite subfamilies $\{(\alpha_i, \beta_i]\}$ of $\{(\alpha_k, \beta_k]\}$. By taking the supremum over all finite subsets of the countably infinite set $\{\ell((\alpha_k, \beta_k])\}$ of positive numbers we get

(a₁) $$\sum_k \ell((\alpha_k, \beta_k]) = \sup \sum_i \ell((\alpha_i, \beta_i]) \leq \ell((a, b]).$$

On the other hand, take an arbitrary $\varepsilon > 0$ and let $\{\varepsilon_k\}$ be any sequence of *positive* numbers such that $\sum \varepsilon_k \leq \varepsilon$. Since $(a, b] = \bigcup_k (\alpha_k, \beta_k]$, it follows that $b \in (\alpha_{k_1}, \beta_{k_1}]$ for some index k_1, and hence $b = \beta_{k_1}$. Moreover, it also follows that $a = \inf_k\{\alpha_k\}$, which ensures the existence of an index $k_0 \neq k_1$ such that $\alpha_{k_0} < a + \varepsilon_0$. Thus the infinite family of *open* intervals

$$\left\{(\alpha_{k_0} - \varepsilon_0, \beta_{k_0} + \varepsilon_0), \, (\alpha_k, \beta_k + \varepsilon_k) \text{ for every } k \neq k_0 \text{ with } k_0 \neq k_1\right\}$$

covers the closed and bounded interval $[a, b]$; that is, it covers the compact interval $[a, b]$ (by the Heine–Borel Theorem). The very definition of compactness says that this family of open intervals has a *finite* subfamily of open intervals that still covers $[a, b]$,

$$\left\{(\alpha_{k_0} - \varepsilon_0, \beta_{k_0} + \varepsilon_0), \, (\alpha_j, \beta_j + \varepsilon_j) \text{ for every } j \in J\right\},$$

where J is a finite index set such that $k_0 \notin J$ and $k_1 \in J$. If the intervals in $\{(\alpha_k, \beta_k]\}$ are pairwise disjoint, then we may assume that $\alpha_{k_0} < \alpha_j < \alpha_{k_1}$ for all $j \in J \backslash \{k_1\}$. Note that J is not empty; at least k_1 is there. Let $n \in \mathbb{N}$ be number of elements of J (i.e., the cardinality of J), and relabel this finite family of open intervals with nonnegative integers $i \in [0, 1, ..., n]$. Since the intervals in $\{(\alpha_k, \beta_k]\}$ are disjoint, we can order the endpoints of the intervals that appear in the above finite family as $\alpha_i < \beta_i \leq \alpha_{i+1} < \beta_{i+1}$ for each $i \in [0, 1, ..., n-1]$, identifying k_0 with 0 and k_1 with n, so that

$$a_0 - \varepsilon_0 < a \leq \alpha_0 < \beta_0 \leq \alpha_i < \beta_i \leq \alpha_{i+1} < \beta_{i+1} < \alpha_n < \beta_n = b < \beta_n + \varepsilon_n$$

for $i \in [1, ..., n-2]$. This finite family of open intervals is then rewritten as

$$\left\{(\alpha_0 - \varepsilon_0, \beta_0 + \varepsilon_0), \, (\alpha_i, \beta_i + \varepsilon_i) \text{ for every } i \in [1, ..., n]\right\},$$

which covers $[a, b]$, thus covering the interval $(a, b]$ of class $\mathcal{C}_1$:

$$(a, b] \subseteq [a, b] \subseteq (\alpha_0 - \varepsilon_0, \beta_0 + \varepsilon_0) \cup \bigcup_{i=1}^{n} (\alpha_i, \beta_i + \varepsilon_i)$$

$$\subseteq (\alpha_0 - \varepsilon_0, \alpha_0] \cup (\alpha_0, \beta_0] \cup (\beta_0, \beta_0 + \varepsilon_0] \cup \bigcup_{i=1}^{n} (\alpha_i, \beta_i] \cup \bigcup_{i=1}^{n} (\beta_i, \beta_i + \varepsilon_i].$$

Take an arbitrary index i in $[0, ..., n{-}1]$. Since $\{(\alpha_k, \beta_k]\}$ consists of disjoint intervals, it follows that $\beta_i \leq \alpha_{i+1}$. Moreover, since the above union of intervals of class $\mathcal{C}_1$ covers $[a, b]$, it also follows that $\alpha_{i+1} \leq \beta_i + \varepsilon_i$. Thus consider a finite sequence $\{\delta_i\}$ of nonnegative numbers given by

$$0 \leq \delta_i = \alpha_{i+1} - \beta_i \leq \varepsilon_i \text{ for each } i \in [0, ..., n{-}1] \quad \text{and} \quad 0 < \delta_n = \varepsilon_n.$$

Replacing $\{\varepsilon_i\}$ with $\{\delta_i\}$ in the intervals of the form $(\beta_i, \beta_i + \varepsilon_i]$ we get a *finite* covering for $(a, b]$ consisting of *disjoint* intervals of class $\mathcal{C}_1$, namely,

$$(\alpha_0 - \varepsilon_0, \alpha_0] \cup (\alpha_0, \beta_0] \cup (\beta_0, \beta_0 + \delta_0] \cup \bigcup_{i=1}^{n} (\alpha_i, \beta_i] \cup \bigcup_{i=1}^{n} (\beta_i, \beta_i + \delta_i].$$

Therefore, according to Definition 8.7(a,c) and Problem 8.4,

$$\ell\big((a, b]\big) \leq \varepsilon_0 + \sum_{i=0}^{n} \ell\big((\alpha_i, \beta_i]\big) + \sum_{i=0}^{n} \delta_i \leq \sum_{k} \ell\big((\alpha_k, \beta_k]\big) + 2\varepsilon$$

for $\sum_{i=0}^{n} \delta_i < \sum_{k} \varepsilon_k \leq \varepsilon$. Because this holds for an arbitrary $\varepsilon > 0$, we get

$$(a_2) \qquad\qquad \ell\big((a, b]\big) \leq \sum_{k} \ell\big((\alpha_k, \beta_k]\big).$$

Thus the result in (a) follows by (a_1) and (a_2).

(b) The disjointness assumption, on both $\{(\alpha_k, \beta_k]\}$ and $\{(a_i, b_i]\}$, ensures that, if $\bigcup_i (a_i, b_i] = \bigcup_k (\alpha_k, \beta_k]$, then $(a_i, b_i] = \bigcup_j (\alpha_{i,j}, \beta_{i,j}]$ for each i, where $\{(\alpha_{i,j}, \beta_{i,j}]\} = \{(\alpha_k, \beta_k]\}$. Therefore, applying Definition 8.7(c) and the result in (a), we get

$$\ell\Big(\bigcup_k (\alpha_k, \beta_k]\Big) = \ell\Big(\bigcup_i (a_i, b_i]\Big) = \sum_i \ell\big((a_i, b_i]\big) = \sum_i \sum_j \ell\big((\alpha_{i,j}, \beta_{i,j}]\big).$$

Since the summands $\{\ell((\alpha_{i,j}, \beta_{i,j}])\}$ are nonnegative (real) numbers, the double infinite sum (which is finite) is unconditionally convergent, and so

$$\sum_i \sum_j \ell\big((\alpha_{i,j}, \beta_{i,j}]\big) = \sum_k \ell\big((\alpha_k, \beta_k]\big). \qquad \square$$

Lemma 8.9. *The collection $\mathfrak{S}$ of all finite unions of intervals from $\mathcal{I}$ is an algebra of subsets of $\mathbb{R}$, and the length function $\ell \colon \mathfrak{S} \to \overline{\mathbb{R}}$ is a measure on the algebra $\mathfrak{S}$.*

Proof. Consider the collection $\Im$ of all finite unions of intervals from $\mathcal{I}$ as in Definition 8.7. It is easy to verify that $\Im$ is an algebra of subsets of $\mathbb{R}$ (see Problem 8.3). We show that ℓ is a measure on that algebra. First note that $\ell(\varnothing) = 0$ by Definition 8.7(a). Now observe that every set in $\Im$ can be expressed as a finite union of *disjoint* intervals from $\mathcal{I}$ so that and $\ell(F) \geq 0$ for every $F \in \Im$ by Definition 8.7(a,b,c). In fact, it is readily verified that every countable family of intervals in $\mathcal{I}$ admits a disjointification *consisting of intervals in $\mathcal{I}$* (see Hints to Problems 2.3 and 7.2). Thus, to complete the proof, it remains to verify axiom (c) of Definition 8.1, namely,

(c) $\ell\left(\bigcup_n F_n\right) = \sum_n \ell(F_n)$

for every countable family $\{F_n\}$ of pairwise disjoint sets in $\Im$ for which $\bigcup_n F_n$ lies in $\Im$.

Since every set in $\Im$ can be expressed as a finite union of *disjoint* intervals from $\mathcal{I}$, it follows that the result in (c) is readily verified by Definition 8.7(c) whenever $\{F_i\}$ is a *finite* family of pairwise disjoint sets in $\Im$. Thus suppose $\{F_n\}$ is a countably infinite family (equivalently, an infinite sequence) of pairwise disjoint sets in $\Im$ such that $\bigcup_n F_n$ lies in $\Im$. Consider the extended nonnegative-valued sequence $\{\ell(F_n)\}$. If $\sum_n \ell(F_n) < \infty$, then $\ell\left(\bigcup_i F_i\right) = \sum_i \ell(F_i) < \sum_n \ell(F_n) < \infty$ for every finite subunion $\bigcup_i F_i$ of $\bigcup_n F_n$ because (c) holds for finite families of disjoint sets in $\Im$. Hence

$$\ell\left(\bigcup_n F_n\right) \leq \sup \ell\left(\bigcup_i F_i\right) \leq \sum_n \ell(F_n) < \infty,$$

where the supremum is taken over all *finite subunions* of the countably infinite union $\bigcup_n F_n$. Therefore, if $\ell\left(\bigcup_n F_n\right) = \infty$, then axiom (c) holds:

$$\ell\left(\bigcup_n F_n\right) = \sum_n \ell(F_n) = \infty.$$

Thus suppose $\ell\left(\bigcup_n F_n\right) < \infty$ so that $\ell(F_n) < \infty$ for each n by Problem 8.4, and recall that each F_n and also $\bigcup_n F_n$ are sets in $\Im$. Then each set F_n is a finite union of disjoint intervals of class $\mathcal{C}_1$, say,

$$F_n = \bigcup_j (\alpha_{n,j}, \beta_{n,j}].$$

Recalling that $\{F_n\}$ is a sequence of disjoint sets, this implies that

$$\bigcup_n F_n = \bigcup_{n,j} (\alpha_{n,j}, \beta_{n,j}] = \bigcup_k (\alpha_k, \beta_k],$$

where $\{(\alpha_{n,j}, \beta_{n,j}]\} = \{(\alpha_k, \beta_k]\}$ is an infinite family of disjoint intervals of class $\mathcal{C}_1$, which also is a finite union of disjoint intervals of class $\mathcal{C}_1$, say,

$$\bigcup_n F_n = \bigcup_i (a_i, b_i].$$

Hence $\bigcup_i (a_i, b_i] = \bigcup_k (\alpha_k, \beta_k]$. Applying Proposition 8.8(b) and recalling the unconditional convergence argument that closed that proof, we get

$$\ell\left(\bigcup_n F_n\right) = \ell\left(\bigcup_i (a_i, b_i]\right) = \sum_k \ell((\alpha_k, \beta_k]) = \sum_{n,j} \ell((\alpha_{n,j}, \beta_{n,j}])$$
$$= \sum_n \sum_j \ell((\alpha_{n,j}, \beta_{n,j}]) = \sum_n \ell\left(\bigcup_j (\alpha_{n,j}, \beta_{n,j}]\right) = \sum_n \ell(F_n)$$

by Definition 8.7(c), so (c) holds, thus completing the proof.   □

Now we are ready to apply the general extension results of Section 8.2 to build up the Lebesgue measure.

Theorem 8.10. *There is a σ-algebra $\mathfrak{S}^*$ of subsets of $\mathbb{R}$ that includes the algebra $\mathfrak{S}$ and extends the length function $\ell \colon \mathfrak{S} \to \overline{\mathbb{R}}$ uniquely to a measure $\lambda^* \colon \mathfrak{S}^* \to \overline{\mathbb{R}}$, which is σ-finite and complete on the σ-algebra $\mathfrak{S}^*$.*

Proof. Since $\ell \colon \mathfrak{S} \to \overline{\mathbb{R}}$ is a measure on the algebra $\mathfrak{S}$ (by Lemma 8.9), let $\ell^* \colon \wp(\mathbb{R}) \to \overline{\mathbb{R}}$ be the outer measure generated by ℓ (as in Definition 8.2), and let $\mathfrak{S}^*$ be the collection of all sets in $\wp(\mathbb{R})$ that are ℓ^*-measurable:

$$\mathfrak{S}^* = \big\{ F \in \wp(\mathbb{R}) \colon \ell^*(F) = \ell^*(S \cap F) + \ell^*(S \backslash F) \text{ for every } S \in \wp(\mathbb{R}) \big\}.$$

Theorem 8.4 ensures that $\mathfrak{S}^*$ is a σ-algebra of subsets of $\mathbb{R}$ that includes the algebra $\mathfrak{S}$, and the restriction of ℓ^* to $\mathfrak{S}^*$ is a measure (i.e., $\mathfrak{S} \subseteq \mathfrak{S}^* \subseteq \wp(\mathbb{R})$ and $\ell^*|_{\mathfrak{S}^*}$ is a measure on $\mathfrak{S}^*$), which, by Proposition 8.3(d), extends ℓ over $\mathfrak{S}^*$. Moreover, it is readily verified that the measure ℓ on the algebra $\mathfrak{S}$ is σ-finite. Indeed, the real line $\mathbb{R}$ is covered by the countably infinite family of intervals $\{(q_k - \varepsilon, q_k + \varepsilon]\}$ of class $\mathcal{C}_1$ of length 2ε for any $\varepsilon > 0$, where $\{q_k\}$ (with k running over all integers $\mathbb{Z}$) is an enumeration of the rational numbers $\mathbb{Q}$ (see Example 2C). Since ℓ is σ-finite, Theorem 8.6 says that there is a unique measure on $\mathfrak{S}^*$, say $\lambda^* \colon \mathfrak{S}^* \to \overline{\mathbb{R}}$, that extends ℓ over $\mathfrak{S}^*$, thus inheriting σ-finiteness from ℓ. By uniqueness, this extension of ℓ over $\mathfrak{S}^*$ coincides with the restriction of ℓ^* to $\mathfrak{S}^*$; that is, $\lambda^* = \ell^*|_{\mathfrak{S}^*}$. Finally, λ^* is a complete measure on the σ-algebra $\mathfrak{S}^*$ according to Proposition 8.5 (i.e., the measure space $(\mathbb{R}, \mathfrak{S}^*, \lambda^*)$ is complete).   □

The σ-algebra $\mathfrak{S}^*$ is referred to as the *Lebesgue algebra*. The sets in $\mathfrak{S}^*$ are called *Lebesgue sets* (or $\mathfrak{S}^*$-measurable) and the measure λ^* on $\mathfrak{S}^*$ is also called *Lebesgue measure*. We close the section by considering a collection (certainly not exhaustive) of basic properties of the Lebesgue measure.

Recall that the Borel algebra $\mathfrak{R}$, consisting of the Borel sets (i.e., $\mathfrak{R}$-measurable sets), is the σ-algebra generated by the open intervals of the real line $\mathbb{R}$ or, equivalently, by the left-open intervals in $\mathcal{I}$, which means that $\Re$ is the intersection of all σ-algebras of subsets of $\mathbb{R}$ that include the family $\mathcal{I}$. Since any σ-algebra that includes the family $\mathcal{I}$ necessarily includes the algebra $\Im$ (finite union of intervals from $\mathcal{I}$), it follows that

(P_1) $\Re$ is the smallest σ-algebra including the algebra $\Im$.

Thus we may conclude that $\Re$ properly includes the algebra $\Im$ (since $\Im$ is not a σ-algebra), and also that $\Re$ is included in the σ-algebra $\Im^*$ (because $\Im^*$ includes $\Im$), which yields the following chain:

$$\Im \subset \Re \subseteq \Im^* \subseteq \wp(\mathbb{R}).$$

Clearly, the restriction of the Lebesgue measure $\lambda \colon \Re \to \overline{\mathbb{R}}$ of Example 2C to the algebra $\Im$ is the length function $\ell \colon \Im \to \overline{\mathbb{R}}$ (cf. Problem 2.7(c)), and the restriction of the outer measure $\ell^* \colon \wp(\mathbb{R}) \to \overline{\mathbb{R}}$ to the Lebesgue algebra $\Im^*$ is the Lebesgue measure $\lambda^* \colon \Im^* \to \overline{\mathbb{R}}$ (cf. Proof of Theorem 8.10):

$$\lambda|_{\Im} = \ell \quad \text{and} \quad \ell^*|_{\Im^*} = \lambda^*.$$

Since the length function ℓ on $\Im$ extends to the measure λ^* on $\Im^*$ so that $\lambda^*|_{\Im} = \ell = \lambda|_{\Im}$, and since this extension is *unique* (Theorem 8.10), we can infer that the restriction of the Lebesgue measure $\lambda^* \colon \Im^* \to \overline{\mathbb{R}}$ to the Borel algebra $\Re$ is precisely the Lebesgue measure $\lambda \colon \Re \to \overline{\mathbb{R}}$ of Example 2C:

(P_2) $\lambda^*|_{\Re} = \lambda.$

Therefore, $\ell = \lambda|_{\Im} = \lambda^*|_{\Im} = \ell^*|_{\Im}$, $\lambda = \lambda^*|_{\Re} = \ell^*|_{\Re}$, and $\lambda^* = \ell^*|_{\Im^*}$. Since λ is a Borel measure in the sense of Problem 2.13, then so is its extension λ^*. Moreover, since $\Re$-measurable sets are $\Im^*$-measurable, all the Borel sets of Problem 2.7 are Lebesgue sets, and their λ^*-measures (as Lebesgue sets) coincide with their λ-measures (as Borel sets), which in turn coincide with their ℓ^*-outer measures. In particular, the Cantor set of Problem 2.9 is an uncountable set with Lebesgue measure zero. The next property says that $(\mathbb{R}, \Im^*, \lambda^*)$ *is the completion of the measure space* $(\mathbb{R}, \Re, \lambda)$. Equivalently, $\Im^*$ *is the completion of the σ-algebra* $\Re$ *with respect to the measure* λ, *or* λ^* *is the completion of the measure* λ *on* $\Re$. Thus

(P_3) $\Im^* = \overline{\Re} \quad \text{and} \quad \lambda^* = \overline{\lambda}$

(cf. Remark that closes Section 2.3). Indeed, since $(\mathbb{R}, \Im^*, \mu^*)$ is a complete measure space, and according to Problem 8.6(b), it follows that

$$\Im^* = \big\{ \overline{E} \subseteq \mathbb{R} : \overline{E} = E \cup A, \ \text{with} \ E \in \Re, \ A \subseteq N \in \Re \ \text{and} \ \lambda(N) = 0 \big\},$$

which is the definition of $\overline{\Re}$. Thus, in light of Problem 2.15, $\Im^*$-*measurable functions are a.e. equal to* $\Re$-*measurable functions*. That is, if $f \colon \mathbb{R} \to \overline{\mathbb{R}}$ is an $\mathfrak{S}^*$-measurable (or *Lebesgue measurable*) function, then there is an $\mathfrak{R}$-measurable (or *Borel measurable*) function $g \colon \mathbb{R} \to \overline{\mathbb{R}}$ such that $g = f$ λ-a.e. It is worth noticing that the Borel–Stieltjes measure on $\mathfrak{R}$ of Example 2D can be naturally extended to a complete *Lebesgue–Stieltjes measure* on $\mathfrak{S}^*$. The above properties are easily verified. However, $\mathfrak{S}^*$ *is neither the smallest nor the largest* σ-*algebra including* $\mathfrak{S}$, so the inclusions below are proper:

$$(\mathrm{P}_4) \qquad\qquad \mathfrak{R} \subset \mathfrak{S}^* \subset \wp(\mathbb{R}).$$

This is all but trivial. It has been a crucial property in all aspects (including historical aspects). Completeness of the measure space $(\mathbb{R}, \mathfrak{S}^*, \lambda^*)$ can be applied to give an existential proof for the first proper inclusion as follows. Let $\mathcal{I}_\mathbb{Q}$ be the subfamily of $\mathcal{I}$ consisting of those intervals from $\mathcal{I}$ with rational endpoints. It is readily verified that this is a countably infinite set $(\#\mathcal{I}_\mathbb{Q} = \#\mathbb{N}$, where $\#$ stands for cardinality), and also that the smallest σ-algebra of subsets of $\mathbb{R}$ that includes $\mathcal{I}_\mathbb{Q}$ is the smallest σ-algebra that includes $\mathcal{I}$, and so it is the smallest σ-algebra that includes the algebra $\mathfrak{S}$, which is precisely the Borel algebra $\mathfrak{R}$. Then $\mathfrak{R}$ is the σ-algebra generated by $\mathcal{I}_\mathbb{Q}$. Thus. since $\#\mathcal{I}_\mathbb{Q} = \#\mathbb{N} \leq \#\mathbb{R}$, it follows that $\#\mathfrak{R} \leq \#\mathbb{R}$ [13, Problem 5.9(c)]. But $\#\mathbb{R} \leq \#\mathfrak{R}$ trivially (for each $x \in \mathbb{R}$, $(x, x+1) \in \mathfrak{R}$). Therefore,

$$\#\mathfrak{R} = \#\mathbb{R}.$$

Now recall that the Cantor set $C \in \mathfrak{R} \subseteq \mathfrak{S}^*$ of Problem 2.9 is uncountable (actually, $\#C = \#\mathbb{R}$) and has measure zero. Since λ^* is a complete measure on $\mathfrak{S}^*$ and $\lambda^*(C) = 0$, it follows that all subsets of C are measurable; that is, $\wp(C) \subseteq \mathfrak{S}^*$. Thus, recalling that $\#X < \#\wp(X)$ for every set X, we get

$$\#\mathbb{R} = \#C < \#\wp(C) \leq \#\mathfrak{S}^*.$$

Hence $\#\mathfrak{R} < \#\mathfrak{S}^*$ and $\mathfrak{R} \subseteq \mathfrak{S}^*$ so that

$$\mathfrak{R} \subset \mathfrak{S}^*.$$

As for the second proper inclusion,

$$\mathfrak{S}^* \subset \wp(\mathbb{R}),$$

a proof is indicated in Problem 8.14, which is based on "translation invariance", a crucial property that we discuss next. But first it is worth noticing that, by Properties P_3 and P_4, the σ-algebra $\mathfrak{R}$ is not complete with respect to λ, and this implies that there are Lebesgue sets of measure zero that are not Borel. However, *every Lebesgue set of measure zero is a subset of a Borel set of measure zero*:

$$(\mathrm{P}_5) \quad N \in \mathfrak{S}^* \ \text{ and } \ \lambda^*(N) = 0 \quad \text{imply} \quad N \subseteq G \in \mathfrak{R} \ \text{ and } \ \lambda(G) = 0.$$

Indeed, since $N \in \mathfrak{I}^*$ and $\lambda^*(N) = 0$ imply $N \in \mathscr{P}(\mathbb{R})$ and $\ell^*(N) = 0$, Property P_5 is straightforward by Problem 8.8(c), which also ensures that the Borel set G of measure zero in fact is a G_δ. We now focus on the *translation invariance* property. For every $\alpha \in \mathbb{R}$ and every $S \subseteq \mathbb{R}$, set

$$S + \alpha = \{\xi + \alpha \in \mathbb{R} \colon \xi \in S\} \subseteq \mathbb{R}.$$

If $E \in \mathfrak{I}^*$, then $E + \alpha \in \mathfrak{I}^*$ and

(P$_6$) $$\lambda^*(E + \alpha) = \lambda^*(E)$$

(cf. Problem 8.12). Translation invariance plays a central role when we set about to build up a *nonmeasurable set* (as we do in Problem 8.14). Another important consequence of it reads as follows.

(P$_7$)   Sets with positive outer measure include nonmeasurable subsets.

(See Problem 8.18.) In particular, *every Lebesgue set of positive measure includes a nonmeasurable set.*

Remarks: Recall the very first paragraph of Chapter 1, where we said that "the power set is too big a domain for a measure". Only now we are ready to offer a proper explanation of that assertion. In fact, it is not possible to construct a set function with the following four properties: (1) defined on the whole $\mathscr{P}(\mathbb{R})$, (2) assigning to each interval the value of its length, (3) countably additive, and (4) translation invariant (reason: (2), (3), and (4) were all we needed in the proof of Problem 8.14). Weakening property (1) is one approach to face the problem that supplies a measure λ^*, defined on a proper σ-algebra $\mathfrak{I}^*$, retaining the useful properties (2), (3), and (4). However, there are other approaches (e.g., if we keep (1), (2), and (4) but replace (3) with subadditivity, then we get the outer measure ℓ^* on $\mathscr{P}(\mathbb{R})$). Assuming the *Continuum Hypothesis* (i.e., the hypothesis that every uncountable subset of $\mathbb{R}$ has the same cardinality as $\mathbb{R}$ itself), then it can be shown that there is no set function satisfying properties (1), (2), and (3) only. (See the references in the Suggested Reading section.)

8.4   Problems

Problem 8.1. As in the case of a measure on a σ-algebra (cf. Definition 2.1), the axiom (c) of Definition 8.1 is referred to as *countable additivity.* Let $\mu \colon \mathcal{A} \to \overline{\mathbb{R}}$ be a set function on an algebra $\mathcal{A}$ and consider the assertion:

  ○ $\mu\left(\bigcup_i E_i\right) = \sum_i \mu(E_i)$ for every finite family $\{E_i\}$ of pairwise disjoint sets in $\mathcal{A}$.

This is referred to as *finite additivity*. Clearly, countable additivity implies finite additivity. First verify that the converse fails. (*Hint:* Problem 2.5.) Now suppose the set function μ is such that

(i) $\mu(E) \geq 0$ for every $E \in \mathcal{A}$

and, if $\{A_n\}$ is a sequence of sets in $\mathcal{A}$,

(ii) $\lim_k \mu\left(\bigcup_{n=1}^k A_n\right) = \mu\left(\bigcup_{n=1}^\infty A_n\right)$

whenever $\bigcup_{n=1}^\infty A_n$ lies in $\mathcal{A}$. Prove the following proposition. *If μ is finitely additive, then it is countably additive.* In other words, take an arbitrary sequence $\{E_n\}$ of pairwise disjoint sets in $\mathcal{A}$. If

$$\mu\left(\bigcup_{n=1}^m E_n\right) = \sum_{n=1}^m \mu(E_n)$$

for every $m \geq 1$ (i.e., *finite additivity*) and (i) and (ii) hold, then show that

$$\mu\left(\bigcup_{n=1}^\infty E_n\right) = \sum_{n=1}^\infty \mu(E_n)$$

whenever $\bigcup_{n=1}^\infty E_n$ lies in $\mathcal{A}$ (i.e., *countable additivity*).

Hint: Follow the proof of Proposition 2.2(a) — recalling that $\mathcal{A}$ is an algebra and μ is finitely additive — to show that: if (i) holds, then

(i') $A, B \in \mathcal{A}$ and $A \subseteq B$ imply $\mu(A) \leq \mu(B)$.

Take an arbitrary sequence $\{E_n\}$ of pairwise disjoint sets in $\mathcal{A}$ such that $\bigcup_{n=1}^\infty E_n \in \mathcal{A}$. Use finite additivity, the fact that $\left\{\bigcup_{n=1}^m E_n\right\}$ is an increasing sequence of sets in $\mathcal{A}$, and the result in (i') to check that

$$\sum_{n=1}^m \mu(E_n) = \mu\left(\bigcup_{n=1}^m E_n\right) \leq \mu\left(\bigcup_{n=1}^\infty E_n\right) \quad \text{for all} \quad m \geq 1.$$

Apply the same argument, recalling from (i') that $\left\{\mu\left(\bigcup_{n=1}^m E_n\right)\right\}$ is an increasing sequence of nonnegative elements from $\overline{\mathbb{R}}$, to verify that

$$\mu\left(\bigcup_{n=1}^k E_n\right) \leq \lim_m \mu\left(\bigcup_{n=1}^m E_n\right) = \lim_m \sum_{n=1}^m \mu(E_n) = \sum_{n=1}^\infty \mu(E_n) \quad \text{for all} \quad k \geq 1.$$

Problem 8.2. Show that properties (a), (b), and (c) in Definition 8.7 actually define the length function $\ell \colon \mathfrak{S} \to \overline{\mathbb{R}}$ on the whole collection $\mathfrak{S}$.

Hint: Every countable family of intervals in $\mathcal{I}$ admits a disjointification consisting of intervals in $\mathcal{I}$ (use the Hints to Problems 2.3 and 7.2). Apply Definition 8.7(c) to show that a length is defined for every set in $\mathfrak{S}$.

Problem 8.3. Consider the setup of Definition 8.7.

(a) Show that $\Im$ is an algebra of subsets of the real line $\mathbb{R}$.

Hint: To verify axiom (b) of Definition 1.1, viz., *the complement of a set in* $\Im$ *belongs to* $\Im$, proceed as follows. Check that (i) if I is an interval in $\mathcal{I}$, then $\mathbb{R}\backslash I$ is the union of no more than two intervals in $\mathcal{I}$ and thus is a set in $\Im$, and (ii) if E and F are sets in $\Im$, then $E \cap F$ is again a set in $\Im$ (since the intersection of any pair of intervals in $\mathcal{I}$ is again an interval in $\mathcal{I}$). Applying the De Morgan laws, $\mathbb{R}\backslash\bigcup_i I_i = \bigcap_i (\mathbb{R}\backslash I_i)$ lies in $\Im$ for every *finite* union $\bigcup_i I_i$ of intervals I_i from $\mathcal{I}$.

(b) But $\Im$ is not a σ-algebra of subsets of the $\mathbb{R}$.

Hint: $\bigcup_{n=1}^{\infty}(2n-1, 2n] = (1, 2] \cup (3, 4] \cup \ldots$ is not a set in $\Im$.

Problem 8.4. Consider the algebra $\Im$ as in Problem 8.3(a). Show that

$$E \subseteq F \quad \text{with} \quad E, F \in \Im \quad \text{implies} \quad \ell(E) \le \ell(F).$$

Hint: $F = E \cup (F\backslash E)$ and $F\backslash E \in \Im$. Recall the Hint to Problem 8.2.

Problem 8.5. Consider the Hint to Problem 8.3(b). It suggests an infinite union in $\Im$ (actually in class $\mathcal{C}_3$) of disjoint intervals of class $\mathcal{C}_1$, namely

$$\bigcup_{k=1}^{\infty}(k, k+1] = (1, 2] \cup (2, 3] \cup (3, 4] \cup \ldots = (1, +\infty),$$

that has an infinite subunion $\bigcup_{n=1}^{\infty}(2n-1, 2n]$ not in $\Im$. Show that this can happen even if the original infinite union is of class $\mathcal{C}_1$ (thus having a finite length) by exhibiting a sequence $\{(\alpha_k, \beta_k]\}$ of intervals of class $\mathcal{C}_1$ such that

$$\bigcup_{k=1}^{\infty}(\alpha_k, \beta_k] = (0, 1] \quad \text{but} \quad \bigcup_{n=1}^{\infty}(\alpha_n, \beta_n] \notin \Im$$

for some subsequence $\{(\alpha_n, \beta_n]\}$ of $\{(\alpha_k, \beta_k]\}$.   *Hint:* $\left\{\left(\frac{1}{2^k}, \frac{1}{2^{k-1}}\right]\right\}$.

Problem 8.6. Consider the Borel and the Lebesgue algebras $\Re \subseteq \Im^*$. Prove the following propositions.

(a) If $F \in \Im^*$, then there exists $E \in \Re$ such that $\lambda^*(F\backslash E) = 0$.

Hint: Recall that $\lambda^*(F) = \lambda^*(F \cap E) + \lambda^*(F\backslash E)$. Thus, if $\lambda^*(F\backslash E) > 0$ for every $E \in \Re$, then $\lambda^*(F \cap E) < \lambda^*(F)$ (whenever $\lambda^*(F) < \infty$) for every $E \in \Re$, which is a contradiction once $\mathbb{R} \in \Re$.

(b) If $F \in \mathfrak{S}^*$, then there exist $E \in \mathfrak{R}$ and $N \in \mathfrak{S}^*$ such that $F = E \cup N$ and $E \cap N = \varnothing$, where $\lambda^*(N) = 0$ and so $\lambda^*(F) = \lambda(E)$.

Problem 8.7. Consider the outer measure $\ell^* \colon \mathcal{P}(\mathbb{R}) \to \overline{\mathbb{R}}$ generated by the length function $\ell \colon \mathfrak{S} \to \mathbb{R}$, which is a measure on the algebra $\mathfrak{S}$ according to Lemma 8.6. Use Definitions 8.2 and 8.7 to show that ℓ^* is given by

(a) $$\ell^*(S) \;=\; \inf_{\{I_n\} \in \mathcal{I}_S} \sum_n \ell(I_n)$$

for every $S \in \mathcal{P}(\mathbb{R})$, where

$$\mathcal{I}_S = \big\{ \{I_n\} \colon I_n \in \mathcal{I} \ \text{ and } \ S \subseteq \textstyle\bigcup_n I_n \big\}$$

is the collection of all countable families $\{I_n\}$ of intervals in $\mathcal{I}$ that cover S. Take any interval I in $\mathcal{I}$ and let I° and I^- denote the interior and closure of I, respectively (with respect to the usual topology of the real line $\mathbb{R}$). Let $\lambda \colon \mathfrak{R} \to \overline{\mathbb{R}}$ be the Lebesgue measure on the σ-algebra $\mathfrak{R}$, which includes the algebra $\mathfrak{S}$ (i.e., $\mathfrak{S} \subset \mathfrak{R}$). Show that both I° and I^- are sets in $\mathfrak{R}$ and

(b) $$\lambda(I^\circ) = \lambda(I) = \lambda(I^-)$$

for every $I \in \mathcal{I}$. (*Hint:* Problem 2.7.) Recall that $I^\circ \subseteq I \subseteq I^-$ and show:

(c) $$\ell^*(S) \;=\; \inf_{\{I_n\} \in \mathcal{I}_S^\circ} \sum_n \lambda(I_n^\circ) \;=\; \inf_{\{I_n\} \in \mathcal{I}_S^-} \sum_n \lambda(I_n^-),$$

where the infimum is taken over all countable coverings of S consisting either of *open* intervals (equivalently, of interior of intervals in $\mathcal{I}$),

$$\mathcal{I}_S^\circ = \big\{ \{I_n\} \colon I_n \in \mathcal{I} \ \text{ and } \ S \subseteq \textstyle\bigcup_n I_n^\circ \big\},$$

or of *closed* intervals (equivalently, of closure of intervals in $\mathcal{I}$),

$$\mathcal{I}_S^- = \big\{ \{I_n\} \colon I_n \in \mathcal{I} \ \text{ and } \ S \subseteq \textstyle\bigcup_n I_n^- \big\}.$$

Hint: The case of covering by closed intervals is readily verified by item (b) since $I \subseteq I^-$ and $\ell(I) = \lambda(I)$. For the case of covering by open intervals, take any $\varepsilon > 0$ and note that for each $I_n \in \mathcal{I}$ there exists an open interval $J_n \in \mathfrak{R}$ such that $I_n \subseteq J_n$ and $\lambda(J_n) \le \ell(I_n) + \frac{\varepsilon}{2^n}$. Thus $\bigcup_n I_n \subseteq \bigcup_n J_n$ and $\sum_n \lambda(J_n) \le \sum_n \ell(I_n) + \varepsilon$ (if n runs over $\mathbb{N}$).

Problem 8.8. Let $\ell^* \colon \mathcal{P}(\mathbb{R}) \to \overline{\mathbb{R}}$ be the outer measure generated by the measure $\ell \colon \mathfrak{S} \to \mathbb{R}$ on the algebra $\mathfrak{S}$, and let $\lambda \colon \mathfrak{R} \to \overline{\mathbb{R}}$ and $\lambda^* \colon \mathfrak{S}^* \to \overline{\mathbb{R}}$ be the Lebesgue measure on the σ-algebras $\mathfrak{R}$ and $\mathfrak{S}^*$, respectively. Recall that $\mathfrak{R} \subseteq \mathfrak{S}^*$ and $\lambda(E) = \lambda^*(E)$ for every $E \in \mathfrak{R}$, and also that open sets are measurable (indeed, open sets are Borel sets by Problem 2.7(d), thus

Lebesgue sets). Take any $S \in \mathcal{P}(\mathbb{R})$ and an arbitrary $\varepsilon > 0$. Show that there exists an open set $U_\varepsilon \subseteq \mathbb{R}$ such that

(a)
$$S \subseteq U_\varepsilon \quad \text{and} \quad \lambda(U_\varepsilon) \le \ell^*(S) + \varepsilon.$$

Thus, since $\ell^*(S) \le \lambda(U_\varepsilon)$ and infimum is the maximum of all lower bounds,

(b)
$$\ell^*(S) = \inf_{U \in \mathcal{T}} \{\lambda(U): S \subseteq U\},$$

where $\mathcal{T}$ is the topology (i.e., the collection of all open sets) of $\mathbb{R}$.

Hint: Consider the result in Problem 8.7(c) with the infimum taken over $\mathcal{I}_S^\circ$. Thus there exists $\{J_n\} \in \mathcal{I}_S^\circ$ such that $\sum_n \lambda(J_n^\circ) \le \ell^*(S) + \varepsilon$. Since every union of open sets is again an open set, put $U = \bigcup_n J_n^\circ$ in $\Re$ and verify that $\lambda(U) = \lambda^*(U) = \ell^*(U) \le \sum_n \ell^*(J_n^\circ) = \sum_n \lambda(J_n^\circ)$ — Proposition 8.3(e).

Every finite intersection of open sets is an open set, but a countable intersection of open sets, which is usually refereed to as a G_δ, is not necessarily an open set, although always measurable. (A G_δ is a Borel set — why?) Show that there exists a G_δ, say $G \in \Re$, such that

(c)
$$S \subseteq G \quad \text{and} \quad \ell^*(S) = \lambda(G).$$

Hint: According to (a) there exists a sequence $\{U_n\}$ of open sets such that $S \subseteq U_n$ and $\lambda(U_n) \le \ell^*(S) + \frac{1}{n}$ for all n. Put $G = \bigcap_n U_n$ in $\Re$ and verify that $S \subseteq G \subseteq U_n$, and hence $\ell^*(S) \le \ell^*(G) = \lambda(G) \le \lambda(U_n)$ for all n.

Problem 8.9. Use Problem 8.8(c) and the fact that λ^* is complete to show that *every set with outer measure zero is a Lebesgue set*:

(a) $N \in \mathcal{P}(\mathbb{R})$ and $\ell^*(N) = 0$ imply $N \in \Im^*$ and $\lambda^*(N) = 0$;

(b) $E, N \in \mathcal{P}(\mathbb{R})$, $E \subseteq N$ and $\ell^*(N) = 0$ imply $E \in \Im^*$ and $\lambda^*(E) = 0$.

If $S \in \mathcal{P}(\mathbb{R})$, then $S \subseteq G$ with $\ell^*(S) = \lambda(G)$ for some $G \in \Re$ by Problem 8.8(c). Since $G = S \cup (G \backslash S)$ and $S = G \backslash (G \backslash S)$, apply item (a) to show the following result.

(c) If $\ell^*(S) < \infty$, then $S \in \Im^*$ if and only if $\ell^*(S \backslash G) = 0$.

Problem 8.10. Consider the setup of Problem 8.8. Show that

$$\lambda^*(E) = \inf_{U \in \Re} \{\lambda(U): U \text{ is open and } E \subseteq U\} \quad \text{for every} \quad E \in \Im^*.$$

Hint: Problem 8.8(b) and Property P_2.

Problem 8.11. Consider again the setup of Problem 8.8. Show that

$$\lambda^*(E) = \sup_{F \in \mathbb{R}} \{\lambda(F) \colon F \text{ is closed and } F \subseteq E\} \quad \text{for every} \quad E \in \mathfrak{S}^*.$$

Hint: Take any $E \in \mathfrak{S}^*$ and an arbitrary $\varepsilon > 0$. Put $E' = \mathbb{R}\backslash E$, the complement of E, which is again a set in $\mathfrak{S}^*$. By Problem 8.8(a), there exists an open set U_ε such that $E' \subseteq U_\varepsilon$ and $\lambda(U_\varepsilon) \leq \lambda^*(E') + \varepsilon$. Now prove that

$$\lambda^*(U_\varepsilon \backslash E') \leq \varepsilon.$$

If $\lambda^*(E') < \infty$, then the above result is immediate by Proposition 2.2(b). If $\lambda^*(E') = \infty$, then proceed as follows. Since λ^* is a σ-finite measure on $\mathfrak{S}^*$, every set in $\mathfrak{S}^*$ is σ-finite. Thus there exists a countable covering of E', say $\{E'_n\}$, made up of measurable subsets of E of finite measure. Since $\lambda^*(E'_n) < \infty$, there exists an open set $U_{\varepsilon,n}$ such that $E'_n \subseteq U_{\varepsilon,n}$ and $\lambda^*(U_{\varepsilon,n} \backslash E'_n) \leq \frac{\varepsilon}{2^n}$. Take the open set $U_\varepsilon = \bigcup_n U_{\varepsilon,n} \subseteq \bigcup_n E'_n = E'$. Show that $\lambda^*(U_\varepsilon \backslash E') \leq \sum_n \lambda^*(U_{\varepsilon,n} \backslash E'_n) \leq \varepsilon$. This proves the displayed inequality. Consider the closed set $F_\varepsilon = \mathbb{R}\backslash U_\varepsilon \subseteq E$. Verify that $E\backslash F_\varepsilon = U_\varepsilon \backslash E'$, and so $\lambda^*(E\backslash F_\varepsilon) = \varepsilon$. Hence $\lambda^*(E) = \lambda^*(E \cap F_\varepsilon) + \lambda^*(E\backslash F_\varepsilon) = \lambda^*(F_\varepsilon) + \varepsilon$. But $\lambda^*(F_\varepsilon) \leq \lambda^*(E)$ and supremum is the minimum of all upper bounds.

Problem 8.12. Prove the translation invariance property (i.e., prove P$_6$).

Hint: Take any α in $\mathbb{R}$. First show that

$$\ell(I + \alpha) = \ell(I)$$

for every $I \in \mathcal{I}$. Now use Problem 8.7(a) to verify that

$$\ell^*(S + \alpha) = \ell^*(S)$$

for every $S \in \mathcal{P}(\mathbb{R})$. Recall that $E \in \mathfrak{S}^*$ if and only if

$$\ell^*(S) = \ell^*(S \cap E) + \ell^*(S\backslash E) \quad \text{for every} \quad S \in \mathcal{P}(\mathbb{R}).$$

Take any A and B in $\mathcal{P}(\mathbb{R})$. Show that $(A + \alpha) \cap B = (A \cap (B - \alpha)) + \alpha$ and $(\mathbb{R}\backslash B) + \alpha = \mathbb{R}\backslash(B + \alpha)$ so that $(A + \alpha)\backslash B = (A\backslash(B - \alpha)) + \alpha$. Therefore, if E lies in $\mathfrak{S}^*$, then verify each of the following steps:

$$\begin{aligned}
\ell^*(S) &= \ell(S - \alpha) = \ell^*\big((S - \alpha) \cap E\big) + \ell^*\big((S - \alpha)\backslash E\big) \\
&= \ell^*\big((S \cap (E + \alpha)) - \alpha\big) + \ell^*\big((S\backslash(E + \alpha)) - \alpha\big) \\
&= \ell^*\big(S \cap (E + \alpha)\big) + \ell^*\big(S\backslash(E + \alpha)\big),
\end{aligned}$$

and so $E + \alpha$ also lies in $\mathfrak{S}^*$. Since $\lambda^* = \ell^*|_{\mathfrak{S}^*}$, conclude the result in P$_6$:

$$E \in \mathfrak{S}^* \quad \text{implies} \quad E + \alpha \in \mathfrak{S}^* \quad \text{and} \quad \lambda^*(E + \alpha) = \lambda^*(E).$$

Problem 8.13. Define a binary operation $\dotplus : [0, 1) \times [0, 1) \to [0, 1)$, called *sum modulo 1*, as follows. If α and β lie in the interval $[0, 1)$, then

$$\alpha \dotplus \beta = \begin{cases} \alpha + \beta, & \alpha + \beta < 1, \\ \alpha + \beta - 1, & \alpha + \beta \geq 1. \end{cases}$$

Now, for every $\alpha \in [0, 1)$ and every $S \subseteq [0, 1)$, set

$$S \dotplus \alpha = \{\xi \dotplus \alpha \in \mathbb{R} : \xi \in S\} \subseteq [0, 1).$$

Prove translation invariance with respect to sum modulo 1.

(a) If $S \subseteq [0, 1)$, then $\ell^*(S \dotplus \alpha) = \ell^*(S)$.

(b) If $E \subseteq [0, 1)$ and $E \in \mathfrak{S}^*$, then $E \dotplus \alpha \in \mathfrak{S}^*$ and $\lambda^*(E \dotplus \alpha) = \lambda^*(E)$.

Hint: Put $E_1 = E \cap [0, 1-\alpha)$ and $E_2 = E \cap [1-\alpha, 1)$, for any $\alpha \in [0, 1)$ and any $E \subseteq [0, 1)$ such that $E \in \mathfrak{S}^*$. Verify that E_i lies in $\mathfrak{S}^*$ for each $i = 1, 2$. Since $\xi + \alpha < 1$ for every $\xi \in E_1$ and $\xi + \alpha \geq 1$ for every $\xi \in E_2$, show that $E_1 \dotplus \alpha = E_1 + \alpha$ and $E_2 \dotplus \alpha = E_2 + (\alpha - 1)$. Thus $E_i \dotplus \alpha$ lies in $\mathfrak{S}^*$ for each $i = 1, 2$. (Why?). Applying translation invariance again (for ordinary sums, as in Problem 8.12), verify each of the following steps:

$$\lambda^*(E \dotplus \alpha) = \lambda^*(E_1 \dotplus \alpha) + \lambda^*(E_2 \dotplus \alpha)$$
$$= \lambda^*(E_1 + \alpha) + \lambda^*(E_2 + \alpha) = \lambda^*(E_1) + \lambda^*(E_2) = \lambda^*(E).$$

Problem 8.14. Consider the relation $\sim$ on the interval $[0, 1)$ defined by

$$\alpha \sim \beta \quad \text{if and only if} \quad \alpha - \beta \in \mathbb{Q}.$$

That is, α and β in $[0, 1)$ are related if their difference is a rational number. This is an equivalence relation. That is, $\sim$ is reflexive (i.e., $\alpha \sim \alpha$), transitive (i.e., $\alpha \sim \beta$ and $\beta \sim \gamma$ imply $\alpha \sim \gamma$), and symmetric (i.e., $\alpha \sim \beta$ implies $\beta \sim \alpha$). Thus $\sim$ induces a partition of $[0, 1)$ into equivalence classes

$$[\alpha] = \{\alpha' \in [0, 1) : \alpha' \sim \alpha\} = \{\alpha' \in [0, 1) : \alpha' - \alpha \in \mathbb{Q}\}.$$

The *Axiom of Choice* ensures the existence of sets consisting of just one element of each equivalence class. These are called *Vitali sets*. Let $V \subseteq [0, 1)$ be a Vitali set and let $\{q_n\}$ be an enumeration of $\mathbb{Q} \cap [0, 1)$. For each n put $V_n = V \dotplus q_n \subseteq [0, 1)$ and show that $\{V_n\}$ forms a partition of $[0, 1)$:

(a) $V_m \cap V_n = \varnothing$ whenever $m \neq n$.

Hint: Suppose $\xi \in V_n \cap V_m$. Show that $\xi = \alpha \dotplus q_n = \beta \dotplus q_m$, with $\alpha, \beta \in V$, $\alpha \dotplus q_n \in V_n$, and $\beta \dotplus q_m \in V_m$. Then $\alpha - \beta \in \mathbb{Q}$ (i.e.,

$\alpha \sim \beta$), so α and β come from the same equivalence class. Since V has only one element from each equivalence class, $\alpha = \beta$. Thus $m = n$.

(b) $\bigcup_n V_n = [0,1)$.

Hint: If $\alpha \in [0,1)$, then α is in some equivalence class (since these classes form a partition of $[0,1)$), and so $\alpha \sim \beta$ for some $\beta \in V \subseteq [0,1)$ (because V has one element from each class), which means $\alpha - \beta \in \mathbb{Q}$ so that $\alpha = \beta + q$ for some q in $\mathbb{Q}$. First verify that, if $\beta \leq \alpha$, then $\alpha = \beta + q_n$ for q_n in $\mathbb{Q} \cap [0,1)$ so that $\alpha = \beta \dotplus q_n$, and hence $\alpha \in V_n$ for some n. Now show that, if $\alpha < \beta$, then $\alpha = \beta - q_m$ for q_m in $\mathbb{Q} \cap (0,1)$ so that $\alpha = \beta \dotplus p_m$ with $p_m = 1 - q_m$ in $\mathbb{Q} \cap (0,1)$, and hence $\alpha \in V_m$ for some m. Thus conclude that $[0,1) \subseteq \bigcup_n V_n$.

Now use Problem 8.13 to prove that $V \notin \mathfrak{S}^*$: *Vitali sets are not measurable.*

Hint: If $V \in \mathfrak{S}^*$, then $\lambda^*([0,1)) = \sum_n \lambda^*(V_n) = \sum_n \lambda^*(V) \neq 1$.

Problem 8.15. Exhibit a sequence $\{V_n\}$ of disjoint sets in $\mathcal{P}(\mathbb{R})$ such that

$$\ell^*\left(\bigcup_n V_n\right) < \sum_n \ell^*(V_n).$$

Problem 8.16. Exhibit a pair of disjoint sets A and B in $\mathcal{P}(\mathbb{R})$ such that

$$\ell^*(A \cup B) \neq \ell^*(A) + \ell^*(B).$$

Hint: The outer measure of the Vitali set V is such that $0 < \ell^*(V) \leq 1$ by Propositions 8.3(c) and 8.5(a). Take the sequence $\{V_n\}$ of Problem 8.14. Put $q_0 = 0$ so that $V_0 = V$. Observe that the equivalence class containing zero is $[0] = \mathbb{Q} \cap [0,1)$. Use the same argument as in the Hint to Problem 8.14(b) and show that $\bigcup_n V_n \backslash V_0 = (0,1)$. Set $A = V_0$ and $B = \bigcup_n V_n \backslash V_0$.

Problem 8.17. Measurable subsets of Vitali sets have measure zero:

$$E \in \mathfrak{S}^* \text{ and } E \subset V \quad \text{imply} \quad \lambda^*(E) = 0.$$

Hint: Consider the setup of Problem 8.14. Put $E_n = E \dotplus q_n \subset V_n$. Show that $\{E_n\}$ is a sequence of disjoint sets in $\mathfrak{S}^*$ with $\lambda^*(E_n) = \lambda^*(E)$ and

$$\sum_n \lambda^*(E_n) = \lambda^*\left(\bigcup_n E_n\right) \leq \ell^*\left(\bigcup_n V_n\right) = \ell^*([0,1)) = 1.$$

Problem 8.18. Prove Property P_7: Take any set $S \in \mathcal{P}(\mathbb{R})$.

If $\ell^*(S) > 0$, then there exists $S_0 \subseteq S$ such that $S_0 \notin \mathfrak{S}^*$.

Hint: If $\ell^*(S) > 0$, then use Problem 8.7(c) to show that there is a translation of it, say S', such that $\ell^*(S' \cap (0,1)) > 0$. Put $A = S' \cap (0,1)$ and $A_n = A \cap V_n$, with $\{V_n\}$ as in Problem 8.14. If all subsets of S are measurable, then show that A_n is measurable. Since $A_n \subseteq V$, Problem 8.17 ensures that $\lambda^*(A_n) = 0$. Thus verify the following contradiction:

$$0 < \ell^*(A) \le \ell^*\left(A \cap \bigcup_n V_n\right) = \ell^*\left(\bigcup_n A_n\right) \le \sum_n \ell^*(A_n) = \sum_n \lambda^*(A_n) = 0.$$

Problem 8.19. Take any $E \in \Im^*$ and any $S \in \wp(\mathbb{R})$. Show that

(a) $\ell^*(E \cup S) + \ell^*(E \cap S) = \lambda^*(E) + \ell^*(S)$,

(b) $E \subseteq S$ and $\lambda^*(E) < \infty$ imply $\ell^*(S \backslash E) = \ell^*(S) - \lambda^*(E)$.

Hint: $\ell^*(E \cup S) = \ell^*((E \cup S) \cap E) + \ell^*((E \cup S) \backslash E) = \lambda^*(E) + \ell^*(S \backslash E)$ and $\ell^*(E \cap S) + \ell^*(S \backslash E) = \ell^*(S)$, proving (a). Replace S with $S \backslash E$ in (a).

Problem 8.20. Take $S \in \wp(\mathbb{R})$ with $\ell^*(S) < \infty$. Show that

(a) $S \in \Im^*$ if and only if $\lambda^*(E) = \ell^*(S)$ for some $E \in \Im^*$ such that $E \subseteq S$,

(b) $S \in \Im^*$ if and only if $\ell^*(S \backslash E) = 0$ for some $E \in \Im^*$ such that $E \subseteq S$.

Hint: For the nontrivial part of (a): $\ell^*(S \backslash E) = \ell^*(S) - \lambda(E) = 0$ by Problem 8.19(b); use Problem 8.9(a) to conclude that $S = (S \backslash E) \cup E \in \Im^*$.

Problem 8.21. Suppose $S \in \wp(\mathbb{R})$ is such that $S \subseteq E$ for some $E \in \Im^*$ with $\lambda^*(E) < \infty$. Show that $S \in \Im^*$ if and only if

$$\lambda^*(E) = \ell^*(S) + \ell^*(E \backslash S).$$

This is a special version of the Carathéodory condition for measurability. It gives a necessary and sufficient condition for a subset of a Lebesgue measurable set of finite measure to be Lebesgue measurable. Prove it.

Hint: If $S \in \Im^*$, then the preceding equation follows from the Carathéodory condition. Conversely, $E \backslash S \subseteq G$ with $\ell^*(E \backslash S) = \lambda(G)$ for some $G \in \Re$ by Problem 8.8(c). Thus show that $\ell^*(E \backslash S) \le \lambda^*(E \cap G) \le \lambda(G) = \ell^*(E \backslash S)$, and so $\lambda^*(E) = \lambda^*(E \cap G) + \lambda^*(E \backslash G) = \ell^*(E \backslash S) + \lambda^*(E \backslash G)$. If the displayed equation holds, then $\lambda^*(E \backslash G) = \ell^*(S) < \infty$, because $\lambda^*(E) < \infty$. Verify that $E \backslash G \subseteq S$, apply Problem 8.20(a), and conclude: $S \in \Im^*$.

Problem 8.22. Prove the following propositions.

(a) Every measurable $E \in \Im^*$ with $0 < \lambda^*(E) < \infty$ has a nonmeasurable partition, A and B in $\wp(\mathbb{R}) \backslash \Im^*$ with $A \cup B = E$ and $A \cap B = \varnothing$, such that

$$\lambda^*(E) = \lambda^*(A \cup B) < \ell^*(A) + \ell^*(B).$$

Hint: By Problem 8.18, there exists a nonmeasurable $A \subset E$. Thus $\{A, B\}$ with $B = E \backslash A$ is a nonmeasurable partition of E so that, by subadditivity, $\lambda^*(E) \leq \ell^*(A) + \ell^*(B)$. Problem 8.21 says that the inequality is strict.

(b) Conversely, take any nonmeasurable set with finite outer measure, say $A \in \wp(\mathbb{R}) \backslash \Im^*$ with $\ell^*(A) < \infty$, and let $G \in \Re$ be such that $A \subset G$ and $\ell^*(A) = \lambda(G)$ as in Problem 8.8(c). Show that $B = G \backslash A \in \wp(\Re) \backslash \Im^*$ is such that $\ell^*(B) > 0$ (by Problem 8.9), and hence

$$\lambda(G) = \lambda(A \cup B) < \ell^*(A) + \ell^*(B).$$

Suggested Reading

Bartle [3], Halmos [13], Royden [22]. See also [18, Part Two].

9
Product

9.1 Product Measure

The *Cartesian product* of two sets X and Y, denoted by $X \times Y$, is the set of all ordered pairs (x, y) where $x \in X$ and $y \in Y$. Suppose $(X, \mathcal{X}, \mu)$ and $(Y, \mathcal{Y}, \nu)$ are measure spaces. We propose in this section to construct a σ-algebra of subsets of $X \times Y$ that is induced by $\mathcal{X}$ and $\mathcal{Y}$ in such a manner as to allow a measure on it to be given by the product of μ and ν. If we consider the product of measures, then we must face the problem of defining the product "zero times infinity", since these are possible values for extended real-valued measures. Thus we declare again that $0 \cdot +\infty = +\infty \cdot 0 = 0$.

Take any $A \subseteq X$ and any $B \subseteq Y$. The Cartesian product $A \times B \subseteq X \times Y$ is called a *a rectangle* from $X \times Y$. If $(X, \mathcal{X})$ and $(Y, \mathcal{Y})$ are measurable spaces, and if E and F are measurable subsets of X and Y, respectively (i.e., if $E \in \mathcal{X}$ and $F \in \mathcal{Y}$), then $E \times F$ is referred to as a *measurable rectangle* from $X \times Y$ (i.e., a rectangle of measurable sets). Let $\mathcal{X} \times \mathcal{Y}$ denote the σ-algebra generated by the measurable rectangles from $X \times Y$ (i.e., the smallest σ-algebra of subsets of $X \times Y$ containing all $E \times F$ with $E \in \mathcal{X}$ and $F \in \mathcal{Y}$), and consider the measurable space $(X \times Y, \mathcal{X} \times \mathcal{Y})$, which is referred to as the *Cartesian product* of the measurable spaces $(X, \mathcal{X})$ and $(Y, \mathcal{Y})$. Obviously, $E \times F$ is a measurable rectangle from $X \times Y$ if and only if $E \times F \in \mathcal{X} \times \mathcal{Y}$.

Example 9A. Let $\mathfrak{R}$ be the Borel algebra of subsets of $\mathbb{R}$ and consider the Cartesian product $(\mathbb{R} \times \mathbb{R}, \mathfrak{R} \times \mathfrak{R})$ of two copies of $(\mathbb{R}, \mathfrak{R})$. Note that open subsets of $\mathbb{R} \times \mathbb{R}$ lie in $\mathfrak{R} \times \mathfrak{R}$. Indeed, $\mathbb{R}^2$ is separable so that open sets of $\mathbb{R}^2$

are countably covered by the topological base consisting of open rectangles made up of open intervals (Problem 1.14). As a matter of fact, it can be shown that the σ-algebra $\Re\times\Re$ coincides with the σ-algebra generated by the open sets of $\mathbb{R}^2$, and hence $\Re\times\Re$ is the *Borel algebra* of subsets of $\mathbb{R}^2$.

Proposition 9.1. *The collection $\mathcal{P}$ of all finite unions of measurable rectangles is an algebra of subsets of $X\times Y$ that is included in $X\times Y$.*

Proof. Every finite union of measurable rectangles admits a disjointification (see Hints to Problems 2.3 and 7.2) *consisting of measurable rectangles,* and hence finite unions of sets in $\mathcal{P}$ remain in $\mathcal{P}$. It is readily verified that the complement of measurable rectangle lies in $\mathcal{P}$, and also that a finite intersection of sets in $\mathcal{P}$ is again a set in $\mathcal{P}$. Using the De Morgan Laws we can infer that the complement of set in $\mathcal{P}$ remains in $\mathcal{P}$. Thus $\mathcal{P}$ is an algebra. Clearly, $\mathcal{P}\subseteq X\times Y$, the σ-algebra generated by measurable rectangles. □

As we just saw, any set in $\mathcal{P}$ admits a disjointification consisting of measurable rectangles, which means that any set in $\mathcal{P}$ can be expressed as a *finite* union of *disjoint* measurable rectangles.

Definition 9.2. Take any set $P = \bigcup_i E_i\times F_i$ in $\mathcal{P}$, where $\{E_i\times F_i\}$ is an arbitrary *finite* partition of P consisting of measurable rectangles from $X\times Y$. Let $\mu\colon X\to\overline{\mathbb{R}}$ and $\nu\colon Y\to\overline{\mathbb{R}}$ be measures on X and Y, respectively, and define a set function $\varpi\colon\mathcal{P}\to\overline{\mathbb{R}}$ by the following *finite* sum: for $P\in\mathcal{P}$,

(a) $\quad \varpi(P) = \varpi\left(\bigcup_i E_i\times F_i\right) = \sum_i \mu(E_i)\,\nu(F_i) = \sum_i \varpi(E_i\times F_i).$

In particular, for each measurable rectangle $E\times F$ from $X\times Y$,

(b) $\qquad\qquad\qquad\qquad \varpi(E\times F) = \mu(E)\,\nu(F).$

Observe that the first identity in (a) just reminds us that $\{E_i\times F_i\}$ is a partition of P, the second being the actual definition of the set function ϖ on $\mathcal{P}$, and the third is a consequence of the second, according to the particular case in (b). By additivity of the measures μ and ν, it is easy to verify that the sums in (a) are "partition invariant": they remain the same for every finite partition of P made up of measurable rectangles. We show that the set function ϖ is a measure on the algebra $\mathcal{P}$, and then we extend this to a measure on the σ-algebra $X\times Y$. But first we need the following auxiliary result to prove that ϖ is countably additive.

Proposition 9.3. *If $\{E_k\times F_k\}$ is a countable family of disjoint rectangles in $X\times Y$ and $A\times B$ also is a rectangle in $X\times Y$, then*

(a) $\quad A\times B = \bigcup_k E_k\times F_k \quad$ *implies* $\quad \varpi(A\times B) = \sum_k \varpi(E_k\times F_k).$

Moreover, if $\{A_i \times B_i\}$ is a finite set of disjoint rectangles in $\mathcal{X} \times \mathcal{Y}$, then

(b) $\displaystyle\bigcup_i A_i \times B_i = \bigcup_k E_k \times F_k$ implies $\displaystyle \varpi\left(\bigcup_i A_i \times B_i\right) = \sum_k \varpi(E_k \times F_k).$

Proof. Take any measurable rectangle $A \times B$ with $A \in \mathcal{X}$ and $B \in \mathcal{Y}$. Let $\{E_k \times F_k\}$ be an arbitrary countable partition of $A \times B$ with E_k in $\mathcal{X}$ and F_k in $\mathcal{Y}$ so that $\{E_k\}$ and $\{F_k\}$ are countable partitions of A and B, respectively. If $\{E_k \times F_k\}$ is a finite partition, then the results in (a) and (b) are trivially verified by Definition 9.2(a). Thus suppose it is countably infinite.

(a) Let $\mathcal{X}_S$ denote that characteristic function of a set S. Then

$$\mathcal{X}_A(x)\,\mathcal{X}_B(y) = \mathcal{X}_{A \times B}\big((x,y)\big) = \sum_k \mathcal{X}_{E_k \times F_k}\big((x,y)\big) = \sum_k \mathcal{X}_{E_k}(x)\,\mathcal{X}_{F_k}(y),$$

for every pair $(x,y) \in X \times Y$. (The second identity holds once $\{E_k \times F_k\}$ is a family of disjoint sets that cover $A \times B$). Fix x, integrate with respect to ν, and apply the Monotone Convergence Theorem (cf. Problem 3.7), to get

$$\mathcal{X}_A(x)\,\nu(B) = \sum_k \mathcal{X}_{E_k}(x)\,\nu(F_k).$$

Now integrate with respect to μ, repeating the same argument, to get

$$\mu(A)\,\nu(B) = \sum_k \mu(E_k)\,\nu(F_k).$$

Hence we get the result in (a) by Definition 9.2(b):

$$\varpi(A \times B) = \sum_k \varpi(E_k \times F_k).$$

(b) The disjointness assumption on both $\{E_k \times F_k\}$ and $\{A_i \times B_i\}$ ensures that, if $\bigcup_i A_i \times B_i = \bigcup_k E_k \times B_k$, then $A_i \times B_i = \bigcup_j E_{i,j} \times F_{i,j}$ for each i, where $\{E_{i,j} \times F_{i,j}\} = \{E_k \times F_k\}$. Therefore, applying Definition 9.2(a) and the result in (a), we get

$$\varpi\left(\bigcup_k E_k \times F_k\right) = \varpi\left(\bigcup_i A_i \times B_i\right) = \sum_i \varpi(A_i \times B_i) = \sum_i \sum_j \varpi(E_{i,j} \times E_{i,j}).$$

If $\varpi(A_i \times B_i) = \infty$ for some i, then $\varpi\left(\bigcup_i A_i \times B_i\right) = \sum_k \varpi(E_k \times F_k) = \infty$ and the result in (b) holds. If $\varpi(A_i \times B_i) < \infty$ for all i, then the summands $\{\varpi(E_{i,j} \times E_{i,j})\}$ are nonnegative (real) numbers by (a), and hence the double infinite sum (which is finite) is unconditionally convergent, so

$$\sum_i \sum_j \varpi(E_{i,j} \times E_{i,j}) = \sum_k \varpi(E_k \times F_k),$$

and the result in (b) still holds. $\qquad\square$

Lemma 9.4. *The set function ϖ is a measure on the algebra $\mathcal{P}$.*

Proof. First note that, $\varpi(\varnothing) = 0$ and $\varpi(P) \geq 0$ for every $P \in \mathcal{P}$, trivially. Take an arbitrary countable family $\{P_n\}$ of pairwise disjoint sets in $\mathcal{P}$ for which $\bigcup_n P_n$ lies in $\mathcal{P}$. Then, as we saw earlier, each set P_n is a finite union of disjoint measurable rectangles, say,

$$P_n = \bigcup_j E_{n,j} \times F_{n,j}.$$

Recalling that $\{P_n\}$ is a sequence of disjoint sets, this implies that

$$\bigcup_n P_n = \bigcup_{n,j} E_{n,j} \times F_{n,j} = \bigcup_k E_k \times F_k,$$

where $\{E_{n,j} \times F_{n,j}\} = \{E_k \times F_k\}$ is a countable family of disjoint measurable rectangles and also a finite union of disjoint measurable rectangles, say,

$$\bigcup_n P_n = \bigcup_i A_i \times B_i.$$

Hence $\bigcup_i A_i \times B_i = \bigcup_k E_k \times F_k$. Applying Proposition 9.3(b) and recalling the unconditional convergence argument that closed that proof we get

$$\varpi\left(\bigcup_n P_n\right) = \varpi\left(\bigcup_i A_i \times B_i\right) = \sum_k \varpi(E_k \times F_k) = \sum_{n,j} \varpi(E_{n,j} \times F_{n,j})$$
$$= \sum_n \sum_j \varpi(E_{n,j} \times F_{n,j}) = \sum_n \varpi\left(\bigcup_j E_{n,j} \times F_{n,j}\right) = \sum_n \varpi(P_n)$$

by Definition 9.2(a). This shows that ϖ is countably additive. $\qquad\square$

Theorem 9.5. (Product Measure Theorem). *If $(X, \mathcal{X}, \mu)$ and $(Y, \mathcal{Y}, \nu)$ are measure spaces, then there exists a measure $\pi \colon \mathcal{X} \times \mathcal{Y} \to \overline{\mathbb{R}}$ such that*

$$\pi(E \times F) = \mu(E) \, \nu(F)$$

for every measurable rectangle $E \times F$ with $E \in \mathcal{X}$ and $F \in \mathcal{Y}$. Moreover, if μ and ν are σ-finite, then such a measure π is unique and also σ-finite.

Proof. Since ϖ is a measure on the algebra $\mathcal{P}$ (Lemma 9.4), it follows by the Carathéodory Extension Theorem (Theorem 8.4) that there exists an extension of ϖ to a measure π^* on an σ-algebra $\mathcal{P}^*$ that includes the algebra $\mathcal{P}$. Moreover, if μ and ν are σ-finite on $\mathcal{X}$ and $\mathcal{Y}$, respectively, then it is readily verified that ϖ is again σ-finite on $\mathcal{P}$, and so the Hahn Extension Theorem (Theorem 8.6) says that π^* is unique on $\mathcal{P}^*$ and clearly σ-finite as well. Now recall that $\mathcal{X} \times \mathcal{Y}$ is the smallest σ-algebra including $\mathcal{P}$. Let π be the restriction of π^* to $\mathcal{X} \times \mathcal{Y}$, so π is a σ-finite measure on $\mathcal{X} \times \mathcal{Y}$ (because π^* is σ-finite), which (by uniqueness of π^*) must be the extension of ϖ over $\mathcal{X} \times \mathcal{Y}$. Summing up:

$$\mathcal{P} \subseteq \mathcal{X} \times \mathcal{Y} \subseteq \mathcal{P}^*, \qquad \pi = \pi^*|_{\mathcal{X} \times \mathcal{Y}}, \quad \text{and} \quad \varpi = \pi|_{\mathcal{P}} = \pi^*|_{\mathcal{P}}.$$

Finally recall from Definition 9.2(b) that $\varpi(E{\times}F) = \mu(E)\,\nu(F)$ for each measurable rectangle $E{\times}F$. Since $\varpi = \pi|_{\mathcal{P}}$, it follows in particular that $\varpi(E{\times}F) = \pi(E{\times}F)$, and hence the measure $\pi\colon \mathcal{X}{\times}\mathcal{Y} \to \mathbb{R}$ is such that

$$\pi(E{\times}F) = \mu(E)\,\nu(F) \quad \text{for every} \quad E{\times}F \in \mathcal{X}{\times}\mathcal{Y}. \qquad \square$$

The value of π at each Cartesian product $E{\times}F$ in $\mathcal{X}{\times}\mathcal{Y}$ is the product of the values of μ and ν at E in $\mathcal{X}$ and F in $\mathcal{Y}$, respectively. This prompts the notation $\pi = \mu{\times}\nu$, which is referred to as the *product measure* (or as the *product* of the measures μ and ν). Accordingly, $(X{\times}Y,\ \mathcal{X}{\times}\mathcal{Y},\ \mu{\times}\nu)$ is the (Cartesian) *product space* of the measure spaces $(X, \mathcal{X}, \mu)$ and $(Y, \mathcal{Y}, \nu)$.

9.2   Sections

Take any subset S of the Cartesian product $X{\times}Y$. For each $x \in X$, the set

$$S_x = \big\{y \in Y\colon (x,y) \in S\big\}$$

is called the *x-section* of S. Similarly, for each $y \in Y$, the set

$$S^y = \big\{x \in X\colon (x,y) \in S\big\}$$

is called and *y-section* of S. The main point in this notation (with subscript and superscript) is to distinguish x-sections (subsets of Y for each $x \in X$) from y-sections (subsets of X for each $y \in Y$). It is worth noticing that sections are not "slices", which means that $S_x \subseteq Y$ (or $S^y \subseteq X$) and $\{x\}{\times}S_x \subseteq X{\times}Y$ (or $S^y{\times}\{y\} \subseteq X{\times}Y$) are, in general, different sets. Also note that sections of a rectangle $A{\times}B \subseteq X{\times}Y$ are either empty or "sides" of the rectangle. That is,

$$(A{\times}B)_x = \begin{cases} B, & x \in A, \\ \varnothing, & x \notin A, \end{cases} \quad \text{and} \quad (A{\times}B)^y = \begin{cases} A, & y \in B, \\ \varnothing, & y \notin B. \end{cases}$$

In particular,

$$(X{\times}Y)_x = Y \quad \text{and} \quad (X{\times}Y)^y = X.$$

Now take any extended real-valued function $f\colon X{\times}Y \to \overline{\mathbb{R}}$ on $X{\times}Y$. For each $x \in X$, the function $f_x\colon Y \to \overline{\mathbb{R}}$ defined by

$$f_x(y) = f(x,y) \quad \text{for every} \quad y \in (X{\times}Y)_x = Y$$

is the *x-section* of f. For each $y \in Y$, the function $f^y\colon X \to \overline{\mathbb{R}}$ defined by

$$f^y(x) = f(x,y) \quad \text{for every} \quad x \in (X \times Y)^y = X$$

is the y-*section* of f. If $f: A \times B \to \overline{\mathbb{R}}$ is defined on a rectangle $A \times B \subseteq X \times Y$, then its x-sections and y-sections are defined on $B \subseteq Y$ and $A \subseteq X$, respectively: $f_x: B \to \overline{\mathbb{R}}$ and $f^y: A \to \overline{\mathbb{R}}$.

Proposition 9.6. *Every section of a measurable set is measurable:*

(a) $E \in \mathcal{X} \times \mathcal{Y}$ *implies* $E^y \in \mathcal{X}$ *and* $E_x \in \mathcal{Y}$ *for every* $x \in X$ *and* $y \in Y$.

Every section of a measurable function is measurable:

(b) *If* $f: X \times Y \to \overline{\mathbb{R}}$ *is* $\mathcal{X} \times \mathcal{Y}$-*measurable, then* $f^y: X \to \overline{\mathbb{R}}$ *is* $\mathcal{X}$-*measurable and* $f_x: Y \to \overline{\mathbb{R}}$ *is* $\mathcal{Y}$-*measurable for every* $x \in X$ *and* $y \in Y$.

Proof. Let $(X, \mathcal{X})$ and $(Y, \mathcal{Y})$ be measurable spaces and let $(X \times Y, \mathcal{X} \times \mathcal{Y})$ be their Cartesian product.

(a) This is a straightforward consequence of Problem 9.5. Indeed, $\mathcal{X} \times \mathcal{Y}$ is included in the collection $(\mathcal{X} \times \mathcal{Y})_X$ of all subsets of $X \times Y$ for which all x-sections are $\mathcal{Y}$-measurable. Therefore, if $E \in \mathcal{X} \times \mathcal{Y}$, then $E_x \in \mathcal{Y}$. Similarly, $E \in \mathcal{X} \times \mathcal{Y}$ also implies $E^y \in \mathcal{X}$. (See Problem 9.5.)

(b) Take an arbitrary $\alpha \in \mathbb{R}$. If a function $f: X \times Y \to \overline{\mathbb{R}}$ is measurable, then the set $\{(x,y) \in X \times Y : f(x,y) > \alpha\}$ is $\mathcal{X} \times \mathcal{Y}$-measurable, and so every x-section $\{(x,y) \in X \times Y : f(x,y) > \alpha\}_x$ is $\mathcal{Y}$-measurable by item (a). Now observe that, for each $x \in \mathcal{X}$,

$$\{(x,y) \in X \times Y : f(x,y) > \alpha\}_x = \{y \in Y : f(x,y) > \alpha\} = \{y \in Y : f_x(y) > \alpha\}.$$

Thus the set $\{y \in Y : f_x(y) > \alpha\}$ is $\mathcal{Y}$-measurable, which means that the function $f_x: Y \to \overline{\mathbb{R}}$ is measurable. Similarly, using the same argument, the function $f^y: X \to \overline{\mathbb{R}}$ is measurable as well. $\square$

We now apply the Monotone Class Lemma (cf. Problems 1.18 and 1.19) to prove a important result that will play a decisive role in the next section.

Lemma 9.7. *If* $(X, \mathcal{X}, \mu)$ *and* $(Y, \mathcal{Y}, \nu)$ *are σ-finite measure spaces, then for each measurable set* E *in* $\mathcal{X} \times \mathcal{Y}$ *the nonnegative functions* $f_E: X \to \overline{\mathbb{R}}$ *and* $g_E: Y \to \overline{\mathbb{R}}$ *defined by*

$$f_E(x) = \nu(E_x) \quad \text{and} \quad g_E(y) = \mu(E^y)$$

for every $x \in X$ *and every* $y \in Y$, *respectively, are measurable functions and*

$$\int_X f_E \, d\mu = \pi(E) = \int_Y g_E \, d\nu.$$

Proof. Let $\mathcal{K}$ be the collection of all sets E in $\mathcal{X} \times \mathcal{Y}$ such that

$$\left\{ f_E \in M(X, \mathcal{X})^+, \ g_E \in M(Y, \mathcal{Y})^+, \ \text{and} \ \int_X f_E \, d\mu = \pi(E) = \int_Y g_E \, d\nu \right\},$$

the subcollection of $\mathcal{X} \times \mathcal{Y}$ for which the conclusion of the lemma holds true. We split the proof into two parts. First we apply the Monotone Class Lemma in part (a) to prove that $\mathcal{K} = \mathcal{X} \times \mathcal{Y}$ for finite measures so that, in this case, the stated result holds true. Then we apply the Monotone Convergence Theorem in part (b) to extend the result to σ-finite measures.

(a) Take any measurable rectangle $A \times B$ in $\mathcal{X} \times \mathcal{Y}$. Recall that x-sections $(A \times B)_x$ are B if $x \in A$ and empty otherwise, and y-sections $(A \times B)^y$ are A if $y \in B$ and empty otherwise. Thus put, for each $x \in X$ and each $y \in Y$,

$$f_{A \times B}(x) = \nu\big((A \times B)_x\big) = \nu(B) \, \chi_A(x),$$

$$g_{A \times B}(y) = \mu\big((A \times B)^y\big) = \mu(A) \, \chi_B(y),$$

which define real-valued functions (if μ and ν are finite) on X and on Y such that (cf. Example 1B and Proposition 1.5) $f_{A \times B} = \nu(B) \chi_A$ is in $M(X, \mathcal{X})^+$, $g_{A \times B} = \mu(A) \chi_B$ is in $M(Y, \mathcal{Y})^+$, and (cf. Problem 3.3.(a))

$$\int_X f_{A \times B} \, d\mu = \nu(B) \mu(A) = \pi(A \times B) = \mu(A) \nu(B) = \int_Y g_{A \times B} \, d\nu.$$

Now let $\mathcal{P}$ be the algebra of Proposition 9.1. Take an arbitrary set P in $\mathcal{P}$. Since $P = \bigcup_{i=1}^n A_i \times B_i$ is a finite union of *disjoint* measurable rectangles $\{A_i \times B_i\}$, the above results ensure that $P \in \mathcal{K}$, and hence $\mathcal{P} \subseteq \mathcal{K}$. In fact,

$$f_P(x) = \nu(P_x) = \nu\big(\big(\bigcup_{i=1}^n A_i \times B_i\big)_x\big) = \nu\big(\bigcup_{i=1}^n (A_i \times B_i)_x\big)$$
$$= \sum_{i=1}^n \nu\big((A_i \times B_i)_x\big) = \sum_{i=1}^n \nu(B_i) \chi_{A_i}(x) = \sum_{i=1}^n f_{A_i \times B_i}(x),$$

$$g_P(y) = \mu(P^y) = \mu\big(\big(\bigcup_{i=1}^n A_i \times B_i\big)^y\big) = \mu\big(\bigcup_{i=1}^n (A_i \times B_i)^y\big)$$
$$= \sum_{i=1}^n \mu\big((A_i \times B_i)^y\big) = \sum_{i=1}^n \mu(A_i) \chi_{B_i}(y) = \sum_{i=1}^n g_{A_i \times B_i}(y),$$

for every $x \in X$ and $y \in Y$, so (cf. Proposition 1.5) $f_P = \sum_{i=1}^n f_{A_i \times B_i}$ is in $M(X, \mathcal{X})^+$, $g_P = \sum_{i=1}^n g_{A_i \times B_i}$ is in $M(Y, \mathcal{Y})^+$, and (cf. Problem 3.7(a))

$$\int_X f_P \, d\mu = \sum_{i=1}^n \int_X f_{A_i \times B_i} \, d\mu$$

$$= \sum_{i=1}^n \mu(A_i) \nu(B_i) = \sum_{i=1}^n \pi(A_i \times B_i) = \pi\Big(\bigcup_{i=1}^n A_i \times B_i\Big) = \pi(P)$$

$$= \sum_{i=1}^n \int_Y g_{A_i \times B_i} \, d\nu = \int_Y g_P \, d\nu.$$

Thus $P \in \mathcal{K}$ and so

(i) $\hspace{4cm} \mathcal{P} \subseteq \mathcal{K}.$

Moreover, let $\{E_n\}$ be an increasing sequence of sets in $\mathcal{K}$. For each n put

$$f_{E_n}(x) = \nu((E_n)_x) \quad \text{and} \quad g_{E_n}(y) = \mu((E_n)^y)$$

for every $x \in X$ and every $y \in Y$. Since $E_n \in \mathcal{K}$, this defines two sequences $\{f_{E_n}\}$ and $\{g_{E_n}\}$ of functions in $\mathcal{M}(X,\mathcal{X})^+$ and in $\mathcal{M}(Y,\mathcal{Y})^+$ such that

$$\int_X f_{E_n}\, d\mu = \pi(E_n) = \int_Y g_{E_n}\, d\nu.$$

Since $\{E_n\}$ is an increasing sequence of sets, $\{f_{E_n}\}$ and $\{g_{E_n}\}$ are increasing sequences of extended real-valued functions so that they converge point-wise. Put $E = \bigcup_n E_n$ in the σ-algebra $\mathcal{X} \times \mathcal{Y}$. Since the sequences $\{(E_n)_x\}$ and $\{(E_n)^y\}$ of sections in $\mathcal{Y}$ and $\mathcal{X}$ are increasing with $\bigcup_n (E_n)_x = E_x$ in $\mathcal{Y}$ and $\bigcup_n (E_n)^y = E^y$ in $\mathcal{X}$ for each $x \in X$ and each $y \in Y$, it follows by Proposition 2.2.(c) that the functions $f_E : X \to \overline{\mathbb{R}}$ and $g_E : Y \to \overline{\mathbb{R}}$ in fact are the pointwise limits of $\{f_{E_n}\}$ and $\{g_{E_n}\}$:

$$\lim_n f_{E_n}(x) = \lim_n \nu((E_n)_x) = \nu\Big(\bigcup_n (E_n)_x\Big) = \nu(E_x) = f_E(x),$$

$$\lim_n g_{E_n}(y) = \lim_n \mu((E_n)^y) = \mu\Big(\bigcup_n (E_n)^y\Big) = \mu(E^y) = g_E(y),$$

for every $x \in X$ and $y \in Y$. Observe, still by Proposition 2.2.(c), that

$$\lim_n \pi(E_n) = \pi\Big(\bigcup_n E_n\Big) = \pi(E),$$

and recall from Proposition 1.8 that $f_E \in \mathcal{M}(X,\mathcal{X})^+$ and $g_E \in \mathcal{M}(Y,\mathcal{Y})^+$. Now apply the Monotone Convergence Theorem (Theorem 3.4) to conclude:

$$\int_X f_E\, d\mu = \lim_n \int_X f_{E_n}\, d\mu = \pi(E) = \lim_n \int_Y g_{E_n}\, d\nu = \int_Y g_E\, d\nu.$$

Hence,

$$E \in \mathcal{K}.$$

On the other hand, let $\{F_n\}$ be a decreasing sequence of sets in $\mathcal{K}$ and put $F = \bigcap_n F_n$ in $\mathcal{X} \times \mathcal{Y}$. Proceeding as before, consider the similarly defined functions f_{F_n} in $\mathcal{M}(X,\mathcal{X})^+$ and g_{F_n} in $\mathcal{M}(Y,\mathcal{Y})^+$ such that

$$\int_X f_{F_n}\, d\mu = \pi(F_n) = \int_Y g_{F_n}\, d\nu$$

for each n. Now suppose that both μ and ν are finite measures, which implies that the product measure $\pi = \mu \times \nu$ is finite as well. Thus we may apply Proposition 2.2(d) for the finite measure π (instead of Proposition 2.2(c)) to verify that $\{f_{F_n}\}$ and $\{g_{F_n}\}$ are both decreasing sequences of real-valued functions so that they converge pointwise, and

$$\lim_n f_{F_n}(x) = f_F(x) \quad \text{and} \quad \lim_n g_{F_n}(y) = g_F(y)$$

for every $x \in X$ and $y \in Y$, which define the real-valued limit functions $f_F \in \mathcal{M}(X, \mathcal{X})^+$ and $g_F \in \mathcal{M}(Y, \mathcal{Y})^+$. Proposition 2.2(d) also ensures that

$$\lim_n \pi(F_n) = \pi\left(\bigcap_n F_n\right) = \pi(F).$$

Since $\{f_{F_n}\}$ and $\{g_{F_n}\}$ are decreasing sequences of nonnegative real-valued measurable functions, and since $\int f_{F_1} \, d\mu$ and $\int g_{F_1} \, d\nu$ are both bounded by $\pi(X \times Y) = \mu(X)\nu(Y)$, which is finite once $\mu(X) < \infty$ and $\nu(Y) < \infty$, it follows by the Dominated Convergence Theorem (Theorem 4.7) that

$$\int_X f_F \, d\mu = \pi(E) = \int_Y g_F \, d\nu.$$

Hence, if the measures μ and ν are finite,

$$F \in \mathcal{K}.$$

Therefore (cf. Problem 1.15), under the finite measure assumption,

(ii) $\mathcal{K}$ is a monotone class.

According to (i) and (ii) the Monotone Class Lemma (see Problems 1.18 and 1.19) ensures that, if μ and ν are finite measures, then

$$\mathcal{K} = \mathcal{X} \times \mathcal{Y}.$$

(b) Now suppose the measures μ and ν are σ-finite so that there exists a pair of *increasing* sequences $\{X_n\}$ and $\{Y_n\}$ of $\mathcal{X}$-measurable and $\mathcal{Y}$-measurable sets covering X and Y, respectively, such that $\mu(X_n) < \infty$ and $\nu(Y_n) < \infty$. For each n consider the σ-algebras $\mathcal{X}_n = \wp(X_n) \cap \mathcal{X}$ and $\mathcal{Y}_n = \wp(Y_n) \cap \mathcal{Y}$ so that μ and ν are finite measures when restricted to them. Take an arbitrary E in $\mathcal{X} \times \mathcal{Y}$. For each n put $E_n = E \cap (X_n \times Y_n)$ in $\mathcal{X}_n \times \mathcal{Y}_n$, so

$$(E_n)_x = \left(E \cap (X_n \times Y_n)\right)_x = E_x \cap (X_n \times Y_n)_x = E_x \cap Y_n \in \mathcal{Y}_n,$$
$$(E_n)^y = \left(E \cap (X_n \times Y_n)\right)^y = E^y \cap (X_n \times Y_n)^y = E^y \cap X_n \in \mathcal{X}_n.$$

Consider the functions $f_{E_n} : X \to \overline{\mathbb{R}}$ and $g_{E_n} : Y \to \overline{\mathbb{R}}$ defined for each n by

$$f_{E_n}(x) = \nu\left((E_n)_x\right) = \nu\left((E_n)_x\right) \chi_{X_n},$$
$$g_{E_n}(x) = \mu\left((E_n)^y\right) = \mu\left((E_n)^y\right) \chi_{Y_n},$$

for every $x \in X$ and $y \in Y$. Let $\mu_n = \mu|_{X_n}$ and $\nu_n = \nu|_{Y_n}$ be the restrictions of μ and ν to $\mathcal{X}_n$ and $\mathcal{Y}_n$ so that $(X_n, \mathcal{X}_n, \mu_n)$ and $(Y_n, \mathcal{Y}_n, \nu_n)$ are finite measure spaces. Since the stated result holds for finite measures (as we saw in (a)), it follows that $f_{E_n} \in \mathcal{M}(X, \mathcal{X})^+$, $g_{E_n} \in \mathcal{M}(Y, \mathcal{Y})^+$, and

$$\int_X f_{E_n}\, d\mu = \int_{X_n} f_{E_n}\, d\mu_n = \pi(E_n) = \int_{Y_n} g_{E_n}\, d\nu_n = \int_Y g_{E_n}\, d\nu.$$

Recall that $\{X_n\}$ and $\{Y_n\}$ are increasing sequences of $\mathcal{X}$-measurable and $\mathcal{Y}$-measurable sets covering X and Y. Thus $\{E_n\}$, $\{(E_n)_x\}$, and $\{(E_n)^y\}$ are increasing sequences of $\mathcal{X} \times \mathcal{Y}$-measurable, $\mathcal{Y}$-measurable and $\mathcal{X}$-measurable sets that cover E, E_x, and E^y, respectively, so $E = \bigcup_n E_n$, $E_x = \bigcup_n (E_n)_x$, and $E^y = \bigcup_n (E_n)^y$. Then Proposition 2.2(c) ensures that $\lim_n \pi(E_n) = \pi(\bigcup_n E_n) = \pi(E)$ and

$$\lim_n f_{E_n}(x) = \lim_n \nu((E_n)_x) = \nu\Big(\bigcup_n (E_n)_x\Big) = \nu(E_x) = f_E(x),$$

$$\lim_n g_{E_n}(y) = \lim_n \mu((E_n)^y) = \mu\Big(\bigcup_n (E_n)^y\Big) = \mu(E^y) = g_E(y),$$

for every $x \in X$ and $y \in Y$. Since $\{f_{E_n}\}$ and $\{g_{E_n}\}$ are increasing sequences (because $\{X_n\}$ and $\{Y_n\}$ are increasing) of nonnegative measurable functions (by (a)), it follows by the Monotone Convergence Theorem (Theorem 3.4) that $f_E \in \mathcal{M}(X, \mathcal{X})^+$, $g_E \in \mathcal{M}(Y, \mathcal{Y})^+$, and

$$\int_X f_E\, d\mu = \lim_n \int_X f_{E_n}\, d\mu = \pi(E) = \lim_n \int_Y g_{E_n}\, d\nu = \int_Y g_E\, d\nu. \qquad \square$$

9.3   The Fubini Theorem

The next two theorems give sufficient conditions for interchanging the order of integration. The first deals with extended real-valued nonnegative functions. The second dismisses nonnegativeness but assumes integrability.

Theorem 9.8. (Tonelli Theorem). *Let* $(X, \mathcal{X}, \mu)$ *and* $(Y, \mathcal{Y}, \nu)$ *be* σ*-finite measure spaces. If* $h \in \mathcal{M}(X \times Y, \mathcal{X} \times \mathcal{Y})^+$, *then the extended real-valued functions* f_h *and* g_h *defined on* X *and* Y *by*

$$f_h(x) = \int_Y h_x\, d\nu \quad \text{and} \quad g_h(y) = \int_X h^y\, d\mu$$

are in $\mathcal{M}(X, \mathcal{X})^+$ *and in* $\mathcal{M}(Y, \mathcal{Y})^+$, *respectively, and*

$$\int_X f_h\, d\mu = \int_{X \times Y} h\, d\pi = \int_Y g_h\, d\nu.$$

Proof. Let $\mathcal{H}$ be the collection of all $h \in \mathcal{M}(X \times Y, \mathcal{X} \times \mathcal{Y})^+$ such that

$$\left\{ f_h \in \mathcal{M}(X, \mathcal{X})^+, \ g_h \in \mathcal{M}(Y, \mathcal{Y})^+, \ \text{and} \ \int_X f_h \, d\mu = \int_{X \times Y} h \, d\pi = \int_Y g_h \, d\nu \right\},$$

the subcollection of $\mathcal{M}(X \times Y, \mathcal{X} \times \mathcal{Y})^+$ for which the conclusion of the theorem holds true. The program is to prove that $\mathcal{H} = \mathcal{M}(X \times Y, \mathcal{X} \times \mathcal{Y})^+$ or, equivalently, $\mathcal{M}(X \times Y, \mathcal{X} \times \mathcal{Y})^+ \subseteq \mathcal{H}$. Let χ_E be the characteristic function of any E in $\mathcal{X} \times \mathcal{Y}$. Recall that χ_E lies in $\mathcal{M}(X \times Y, \mathcal{X} \times \mathcal{Y})^+$ (Example 1B) and note that $(\chi_E)_x = \chi_{E_x}$ and $(\chi_E)^y = \chi_{E^y}$ (Problem 9.8). Put

$$f_{\chi_E}(x) = \int_Y (\chi_E)_x \, d\nu = \int_Y \chi_{E_x} \, d\nu = \nu(E_x)$$

and

$$g_{\chi_E}(y) = \int_X (\chi_E)^y \, d\mu = \int_X \chi_{E^y} \, d\mu = \mu(E^y)$$

for every $x \in X$ and $y \in Y$. Since $(X, \mathcal{X}, \mu)$ and $(Y, \mathcal{Y}, \nu)$ are σ-finite measure spaces, Lemma 9.7 ensures that $f_{\chi_E} \in \mathcal{M}(X, \mathcal{X})^+$, $g_{\chi_E} \in \mathcal{M}(Y, \mathcal{Y})^+$, and

$$\int_X f_{\chi_E} \, d\mu = \int_{X \times Y} \chi_E \, d\pi = \int_Y g_{\chi_E} \, d\nu,$$

which shows that every characteristic function of sets in $\mathcal{X} \times \mathcal{Y}$ lies in $\mathcal{H}$. Thus, by linearity of the space of measurable functions (Proposition 1.9) and of the integral itself (Proposition 3.5(a,b)), we can infer that every simple function (cf. Definition 3.1) lies in $\mathcal{H}$. Now take an arbitrary function h in $\mathcal{M}(X \times Y, \mathcal{X} \times \mathcal{Y})^+$. According to Problem 1.6, there exists an increasing sequence $\{\varphi_n\}$ of simple functions in $\mathcal{M}(X \times Y, \mathcal{X} \times \mathcal{Y})^+$ converging pointwise to h. For each n and for every $x \in X$ and $y \in Y$, put

$$f_{\varphi_n}(x) = \int_Y (\varphi_n)_x \, d\nu \quad \text{and} \quad g_{\varphi_n}(y) = \int_X (\varphi_n)^y \, d\mu. \qquad (*)$$

Since $\varphi_n \in \mathcal{H}$ (because simple functions are in $\mathcal{H}$), this defines a pair of functions f_{φ_n} and g_{φ_n} such that $f_{\varphi_n} \in \mathcal{M}(X, \mathcal{X})^+$, $g_{\varphi_n} \in \mathcal{M}(Y, \mathcal{Y})^+$, and

$$\int_X f_{\varphi_n} \, d\mu = \int_{X \times Y} \varphi_n \, d\pi = \int_Y g_{\varphi_n} \, d\nu. \qquad (**)$$

Since $\varphi_n \to h$, it is readily verified that $(\varphi_n)_x \to h_x$ and $(\varphi_n)^y \to h^y$ for every $x \in X$ and $y \in Y$, where the foregoing convergences are all pointwise. Moreover, since $\{\varphi_n\}$ is an increasing sequence of nonnegative functions, it is clear that the sequences $\{(\varphi_n)_x\}$ and $\{(\varphi_n)^y\}$ also are increasing and consist of nonnegative functions for each $x \in X$ and $y \in Y$. Furthermore, since φ_n is measurable, then so are all sections $(\varphi_n)_x$ and $(\varphi_n)^y$, for each n, by Proposition 9.6. Therefore, applying the Monotone Convergence Theorem

(Theorem 3.4) we get from (*) that

$$\lim_n f_{\varphi_n}(x) = \lim_n \int_Y (\varphi_n)_x \, d\nu = \int_Y \lim_n (\varphi_n)_x \, d\nu = \int_Y h_x \, d\nu = f_h(x),$$

$$\lim_n g_{\varphi_n}(y) = \lim_n \int_X (\varphi_n)^y \, d\mu = \int_X \lim_n (\varphi_n)^y \, d\mu = \int_X h^y \, d\mu = g_h(y),$$

for every $x \in X$ and every $y \in Y$, so $f_{\varphi_n} \to f_h$ and $g_{\varphi_n} \to g_h$ pointwise. Observe that $\{f_{\varphi_n}\}$ and $\{g_{\varphi_n}\}$ are increasing sequences (since $\{\varphi_n\}$ is increasing). Applying the Monotone Convergence Theorem again we get from (**) that $f_h \in \mathcal{M}(X, \mathcal{X})^+$, $g_h \in \mathcal{M}(Y, \mathcal{Y})^+$, and

$$\int_X f_h \, d\mu = \int_{X \times Y} h \, d\pi = \int_Y g_h \, d\nu,$$

so $h \in \mathcal{H}$. Then $\mathcal{M}(X \times Y, \mathcal{X} \times \mathcal{Y})^+ \subseteq \mathcal{H}$.   □

Note that the conclusion of the Tonelli Theorem can be rewritten as

$$\int_X \left(\int_Y h \, d\nu \right) d\mu = \int_{X \times Y} h \, d\pi = \int_Y \left(\int_X h \, d\mu \right) d\nu,$$

which shows that the order of the integrals can be interchanged. The same applies to the next theorem, which has exactly the same conclusion but a different hypothesis. There the function h is not necessarily nonnegative but should be integrable with respect to the product measure.

Theorem 9.9. (Fubini Theorem) *Let* $(X, \mathcal{X}, \mu)$ *and* $(Y, \mathcal{Y}, \nu)$ *be σ-finite measure spaces. If* $h \in \mathcal{L}(X \times Y, \mathcal{X} \times \mathcal{Y}, \pi)$, *then there are real-valued functions* f_h *and* g_h *defined almost everywhere on* X *and* Y *by*

$$f_h(x) = \int_Y h_x \, d\nu \quad \text{and} \quad g_h(y) = \int_X h^y \, d\mu,$$

which lie in $\mathcal{L}(X, \mathcal{X}, \mu)$ *and in* $\mathcal{L}(Y, \mathcal{Y}, \nu)$, *respectively, such that*

$$\int_X f_h \, d\mu = \int_{X \times Y} h \, d\pi = \int_Y g_h \, d\nu.$$

Proof. Take any h in $\mathcal{M}(X \times Y, \mathcal{X} \times \mathcal{Y})$. Consider its positive and negative parts h^+ and h^- in $\mathcal{M}(X \times Y, \mathcal{X} \times \mathcal{Y})^+$ such that $h = h^+ - h^-$ (cf. Proposition 1.6), and its x-section h_x in $\mathcal{M}(Y, \mathcal{Y})$ and y-section h^y in $\mathcal{M}(X, \mathcal{X})$ (cf. Proposition 9.6). Also consider the parts of the sections $(h_x)^\pm$ in $\mathcal{M}(Y, \mathcal{Y})^+$ and $(h^y)^\pm$ in $\mathcal{M}(X, \mathcal{X})^+$ that, according to Problem 9.9, coincide with the sections of the parts $(h^\pm)_x$ and $(h^\pm)^y$. Since the positive and negative parts h^+ and h^- lie in $\mathcal{M}(X \times Y, \mathcal{X} \times \mathcal{Y})^+$, since $(h^\pm)_x = (h_x)^\pm$ and

$(h^\pm)^y = (h^y)^\pm$, and since the measure spaces are σ-finite, it follows by the Tonelli Theorem that the functions $f_{h\pm}$ and $g_{h\pm}$ defined on X and Y by

$$f_{h\pm}(x) = \int_Y (h_x)^\pm \, d\nu \quad \text{and} \quad g_{h\pm}(y) = \int_X (h^y)^\pm \, d\mu \qquad (*)$$

are in $\mathcal{M}(X, \mathcal{X})^+$ and in $\mathcal{M}(Y, \mathcal{Y})^+$, respectively, and

$$\int_X f_{h\pm} \, d\mu = \int_{X \times Y} h^\pm \, d\pi = \int_Y g_{h\pm} \, d\nu. \qquad (**)$$

Now suppose $h \in \mathcal{L}(X \times Y, \mathcal{X} \times \mathcal{Y}, \pi)$ so that the parts h^+ and h^- are non-negative functions in $\mathcal{L}(X \times Y, \mathcal{X} \times \mathcal{Y}, \pi)$ by Definition 4.1. Although their sections are all real-valued, the nonnegative measurable functions $f_{h\pm}$ and $g_{h\pm}$ are not necessarily real-valued (see Problem 9.11) but have finite integrals according to $(**)$. Then they are real-valued almost everywhere by Problem 3.9(b) and the differences $f_{h+}(x) - f_{h-}(x)$ and $g_{h+}(y) - g_{h-}(y)$ are defined almost everywhere with respect to μ and ν. Thus let f_h and g_h be real-valued functions defined almost everywhere on X and Y by

$$f_h = f_{h+} - f_{h-} \quad \text{and} \quad g_h = g_{h+} - g_{h-}$$

and zero otherwise, which lie in $\mathcal{L}(X, \mathcal{X}, \mu)$ and $\mathcal{L}(Y, \mathcal{Y}, \nu)$. Indeed (cf. Problems 3.8 an 3.9), there is an $\mathcal{X}$-measurable set N with $\mu(N) = 0$ such that $\int_X f_{h\pm}|_{X \setminus N} \, d\mu = \int_X f_{h\pm} \chi_{X \setminus N} \, d\mu = \int_{X \setminus N} f_{h\pm} \, d\mu = \int_X f_{h\pm} \, d\mu < \infty$ for which $f_{h\pm}|_{X \setminus N}$ is *real-valued*, so $f_{h\pm}|_{X \setminus N}$ lies in $\mathcal{L}(X, \mathcal{X}, \mu)$. Hence the function f_h defined by $f_{h+}|_{X \setminus N} - f_{h-}|_{X \setminus N}$ on $X \setminus N$ and zero on N is real-valued and lies in $\mathcal{L}(X, \mathcal{X}, \mu)$ — see Lemma 4.5. Similarly, the function g_h is real-valued and lies in $\mathcal{L}(Y, \mathcal{Y}, \nu)$. Therefore, by $(*)$ and Definition 4.1,

$$f_h(x) = \int_Y (h_x)^+ \, d\nu - \int_Y (h_x)^- \, d\nu = \int_Y h_x \, d\nu$$

and

$$g_h(y) = \int_X (h^y)^+ \, d\nu - \int_X (h^y)^- \, d\nu = \int_X h^y \, d\mu$$

for μ-almost every $x \in X$ and ν-almost every $y \in Y$. Moreover, by $(**)$ and using Definition 4.1 again, and recalling that $f_h = f_{h+} - f_{h-}$ on $X \setminus N$ so that $\int_X f_h \, d\mu = \int_{X \setminus N} f_{h-} \, d\mu - \int_{X \setminus N} f_{h-} \, d\mu$ — see Proposition 4.3, we get

$$\int_{X \times Y} h \, d\pi = \int_{X \times Y} h^+ \, d\pi - \int_{X \times Y} h^- \, d\pi = \int_X f_{h+} \, d\mu - \int_X f_{h-} \, d\mu = \int_X f_h \, d\mu,$$

and also

$$\int_{X \times Y} h \, d\pi = \int_Y g_{h+} \, d\mu - \int_Y g_{h-} \, d\mu = \int_Y g_h \, d\mu. \qquad \square$$

Let N be a measurable set with $\mu(N) = 0$ for which $f_h(x) = \int_Y h_x(y)d\nu$ on $X \backslash N$ and zero on N, as we saw in the preceding proof. Now observe that $\int_X f_h(x)d\mu = \int_{X \backslash N} f_h(x)d\mu = \int_{X \backslash N} \left(\int_Y h_x(y)d\nu \right)d\mu = \int_X \left(\int_Y h(x,y)d\nu \right)d\mu$. Similarly, $\int_Y g_h(y)d\nu = \int_Y \left(\int_X h(x,y)d\mu \right)d\nu$. Therefore, as we commented before, the conclusion of the Fubini Theorem can also be rewritten as

$$\int_X \left(\int_Y h\,d\nu \right)d\mu = \int_{X \times Y} h\,d\pi = \int_Y \left(\int_X h\,d\mu \right)d\nu.$$

The middle integral, $\int_{X \times Y} h\,d\pi$, is the *double integral* of h. The left and right integrals, $\int_X \left(\int_Y h\,d\nu \right)d\mu$ and $\int_Y \left(\int_X h\,d\mu \right)d\nu$, are the *iterated integrals* of h.

9.4   Problems

Problem 9.1. Consider the Borel algebra $\Re \times \Re$ of subsets of $\mathbb{R}^2$ as in Example 9A. If $E \in \Re$, then show that

$$D_E = \left\{ (x,y) \in \mathbb{R}^2 \colon y - x \in E \right\} \in \Re \times \Re.$$

Hint: (a) The "smart" way: $f(x,y) = y - x$ defines a continuous function f from $\mathbb{R}^2$ to $\mathbb{R}$, thus a measurable function, and $D_E = f^{-1}(E)$.

(b) The "tour de force" way: Take an arbitrary E in $\Re$. If E is unbounded, then consider a countable partition of it made up of bounded measurable sets. If E is bounded, then rewrite it as a countable union of *measurable triangles* as follows. With diam E standing for the diameter of E, set

$$E_k = (E \cup \{\sup E\}) + k \operatorname{diam} E$$

and

$$F_k = \left[(E \cup \{\sup E\}) + \inf E \right] + k \operatorname{diam} E$$

for each integer $k \in \mathbb{Z}$, put

$$L = \left\{ (x,y) \in \mathbb{R}^2 \colon y \le \sup E + x \right\} \quad \text{or} \quad L = \left\{ (x,y) \in \mathbb{R}^2 \colon y < \sup E + x \right\}$$

whether $\sup E$ lies or does not lie in E,

$$U = \left\{ (x,y) \in \mathbb{R}^2 \colon y \ge \inf E + x \right\} \quad \text{or} \quad U = \left\{ (x,y) \in \mathbb{R}^2 \colon y > \inf E + x \right\}$$

whether $\inf E$ lies or does not lie in E, and

$$\Delta_k = (E_k \times F_k) \cap U \quad \text{and} \quad \nabla_k = (E_k \times F_{k+1}) \cap L$$

so that the above *triangles* cover the "strip-shape" set D_E (draw a picture):

$$D_E = \left\{ (x,y) \in \mathbb{R}^2 \colon y - x \in E \right\} = \bigcup_k (\Delta_k \cup \nabla_k).$$

Problem 9.2. If $f: \mathbb{R} \to \mathbb{R}$ is $\mathfrak{R}$-measurable, then show that $h: \mathbb{R}^2 \to \mathbb{R}$ given by $h(x, y) = f(y - x)$ for every (x, y) in $\mathbb{R}^2$ is $\mathfrak{R} \times \mathfrak{R}$-measurable.

Hint: Problem 9.1.

Problem 9.3. Take the Cartesian product $(X \times Y, \mathcal{X} \times \mathcal{Y})$ of the measurable spaces $(X, \mathcal{X})$ and $(Y, \mathcal{Y})$. If $f \in \mathcal{M}(X, \mathcal{X})$ and $g \in \mathcal{M}(Y, \mathcal{Y})$ are real-valued functions, then the real-valued function h on $X \times Y$ defined by $h(x, y) = f(x) \, g(y)$ for every (x, y) in $X \times Y$ lies in $\mathcal{M}(X \times Y, \mathcal{X} \times \mathcal{Y})$. Prove.

Problem 9.4. Let S be an arbitrary subset of the Cartesian product $X \times Y$ of two sets X and Y, let $\{S_\alpha\}$ be any collection of subsets of $X \times Y$, and take any x in X. Use the definition of an x-section to show that

(a) $((X \times Y) \backslash S))_x = (X \times Y)_x \backslash S_x = Y \backslash S_x$,

(b) $(\bigcup_\alpha S_\alpha)_x = \bigcup_\alpha (S_\alpha)_x$.

Problem 9.5. Take two measurable spaces $(X, \mathcal{X})$ and $(Y, \mathcal{Y})$ and let the measurable space $(X \times Y, \mathcal{X} \times \mathcal{Y})$ be their Cartesian product. Consider the collections of all subsets of $X \times Y$ for which all x-sections are $\mathcal{Y}$-measurable,

$$(\mathcal{X} \times \mathcal{Y})_X = \{E \subseteq X \times Y : E_x \in \mathcal{Y} \text{ for every } x \in X\},$$

and of all subsets of $X \times Y$ for which all y-sections are $\mathcal{X}$-measurable,

$$(\mathcal{X} \times \mathcal{Y})_Y = \{E \subseteq X \times Y : E^y \in \mathcal{X} \text{ for every } y \in Y\}.$$

Use Problem 9.4 to verify that $(\mathcal{X} \times \mathcal{Y})_X$ and $(\mathcal{X} \times \mathcal{Y})_Y$ are σ-algebras of subsets of $X \times Y$. Show that both σ-algebras $(\mathcal{X} \times \mathcal{Y})_X$ and $(\mathcal{X} \times \mathcal{Y})_Y$ contain all measurable rectangles (recall: sections of a rectangle are either empty or a side of the rectangle), and conclude that

$$\mathcal{X} \times \mathcal{Y} \subseteq (\mathcal{X} \times \mathcal{Y})_X \quad \text{and} \quad \mathcal{X} \times \mathcal{Y} \subseteq (\mathcal{X} \times \mathcal{Y})_Y.$$

Problem 9.6. Take the unit interval $X = [0, 1] \subset \mathbb{R}$ and let $\mathcal{X}$ be the collection of all subsets of X that either are countable or are the complement of a countable set. Verify that $\mathcal{X}$ is a σ-algebra of subsets of X and take the Cartesian product $(X \times X, \mathcal{X} \times \mathcal{X})$ of two copies of the measurable space $(X, \mathcal{X})$. For each $\alpha > 0$ consider the *line segments*

$$D_\alpha = \{(x, y) \in X \times X : y = \alpha x\}.$$

Show that every D_α is nonmeasurable (i.e., $D_\alpha \notin \mathcal{X} \times \mathcal{X}$ for every $\alpha > 0$).

Hint: Let $D_0 = \{(x,y) \in X \times X : y = 0\} = [0,1] \times \{0\}$ be the horizontal line segment. Take an arbitrary line segment D_α distinct from D_0 and consider the intersection of their complements, which is the *sector*

$$(X \backslash D_0) \cap (X \backslash D_\alpha) = \{(x,y) \in X \times X : 0 < y < \alpha x\}.$$

Use Proposition 9.6(a) (or Problem 9.5) to show that such a sector is not $\mathcal{X} \times \mathcal{X}$-measurable. Verify that D_0 is $\mathcal{X} \times \mathcal{X}$-measurable. If D_α is $\mathcal{X} \times \mathcal{X}$-measurable, then so is the intersection of their complements (isn't it?), which is a contradiction. Thus D_α is not $\mathcal{X} \times \mathcal{X}$-measurable.

Problem 9.7. Apply Proposition 9.6(a) to prove the failure of its own converse: *There exist nonmeasurable sets for which all sections are measurable.*

Hint: Problem 9.6.

Problem 9.8. Let $\mathcal{X}_S$ be the characteristic function of a subset S of the Cartesian product $X \times Y$ of two sets X and Y. Show that, for each $x \in X$ and each $y \in Y$, the sections $(\mathcal{X}_S)_x$ and $(\mathcal{X}_S)^y$ of the function $\mathcal{X}_S$ are the characteristic functions of the sections S_x and S^y of the set S. That is,

$$(\mathcal{X}_S)_x = \mathcal{X}_{S_x} \quad \text{and} \quad (\mathcal{X}_S)^y = \mathcal{X}_{S^y}.$$

Thus conclude that, if $A \times B$ is any rectangle from $X \times Y$, then

$$\mathcal{X}_{A \times B}(x,y) = \mathcal{X}_A(x) \mathcal{X}_B(y) \quad \text{for every} \quad (x,y) \in X \times Y.$$

Problem 9.9. Take any extended real-valued function $f : X \times Y \to \overline{\mathbb{R}}$ on the Cartesian Product. Consider its positive and negative parts, f^+ and f^-, and also its x-sections and y-sections, f_x and f^y. Show that

$$(f^+)_x = (f_x)^+, \quad (f^-)_x = (f_x)^- \quad \text{and} \quad (f^+)^y = (f^y)^+, \quad (f^-)^y = (f^y)^-.$$

Hint: Recall that $f^+ = f\mathcal{X}_{F^+}$, where $F^+ = \{(x,y) \in X \times Y : f(x,y) \geq 0\}$, and verify that $(f^+)_x = (f\mathcal{X}_{F^+})_x = f_x(\mathcal{X}_{F^+})_x = f_x\mathcal{X}_{F^+} = (f_x)^+$.

Now show that the x-sections and y-sections of $f = f^+ - f^-$ are given by

$$f_x = (f^+)_x - (f^-)_x \quad \text{and} \quad f^y = (f^+)^y - (f^-)^y.$$

Problem 9.10. Take the Borel algebra $\Re$ (or the Lebesgue algebra $\Im^*$) and the Lebesgue measure λ on $\Re$ (or on $\Im^*$), which is sometimes also referred to as the *linear Lebesgue measure* or even as the *length* on $\mathbb{R}$. Consider the product of λ with itself, that is, the measure $\pi = \lambda \times \lambda$ on the σ-algebra $\Re \times \Re$ (or on $\Im^* \times \Im^*$) of subsets of the plane $\mathbb{R}^2 = \mathbb{R} \times \mathbb{R}$. This measure is referred to as the *planar Lebesgue measure* (or simply as the *area* on $\mathbb{R}^2$). Note that $\mathfrak{R} \times \mathfrak{R} \subset \mathfrak{S}^* \times \mathfrak{S}^*$, so $\mathfrak{R} \times \mathfrak{R}$-measurable sets are $\mathfrak{S}^* \times \mathfrak{S}^*$-measurable. Thus the area of a measurable rectangle $E \times F$ from $\mathbb{R}^2$ is $\lambda(E)\lambda(F)$. Give an example of an uncountable measurable subset of the rectangle $[0,1] \times [0,1]$ with area zero and such that all sections of it (x-sections and y-sections) either are empty or are uncountable with length zero.

Problem 9.11. Consider the Lebesgue measure space $(\mathbb{R}, \mathfrak{R}, \lambda)$ and give an example of a real-valued function $f: \mathbb{R}^2 \to \mathbb{R}$ with the following properties.

(1) $f = 0$ π-almost everywhere.

That is, with respect to the product space $(\mathbb{R}^2, \mathfrak{R} \times \mathfrak{R}, \pi)$, where $\pi = \lambda \times \lambda$ is the area on $\mathbb{R}^2$ (cf. Problem 9.10), $f = 0$ up to a measurable set of area zero, so f is measurable (why?) and also integrable with $\int_{\mathbb{R}^2} f \, d\pi = 0$ according to Proposition 4.2(a). Moreover,

(2) $\int_{\mathbb{R}} f_x \, d\lambda = \infty$ for some x-section $f_x: \mathbb{R} \to \mathbb{R}$ of f.

Problem 9.12. Put $E = [0,1] \subset \mathbb{R}$ and let $\mathcal{E} = \wp(E) \cap \mathfrak{R}$, the σ-algebra of all Borel subsets of $[0,1]$. Let λ be the Lebesgue measure on $\mathcal{E}$ (i.e., the restriction of the Lebesgue measure to $\mathcal{E}$ as in Problem 2.11) and let μ be the counting measure on $\mathcal{E}$ (as in Problem 2.4(b), which is not σ-finite). Consider the measure spaces $(E, \mathcal{E}, \lambda)$ and $(E, \mathcal{E}, \mu)$, use Example 9A to verify that $D = \{(x,y) \in E \times E: x = y\}$ is $\mathcal{E} \times \mathcal{E}$-measurable, and show that

$$\int \mu(D_x) \, d\lambda \neq \int \lambda(D^y) \, d\mu.$$

Thus the assumption of σ-finiteness cannot be omitted from Lemma 9.7.

Problem 9.13. Consider the setup of the previous problem. Use the characteristic function $\chi_D \in \mathcal{M}(E \times E, \mathcal{E} \times \mathcal{E}, \lambda \times \nu)^+$ of the *diagonal subset* D of the rectangle $[0,1] \times [0,1]$ to show that the assumption of σ-finiteness cannot be omitted from the Tonelli Theorem.

Problem 9.14. Let $(X, \mathcal{X}, \mu)$ and $(Y, \mathcal{Y}, \nu)$ be σ-finite measure spaces and let $(X \times Y, \mathcal{X} \times \mathcal{Y}, \pi)$ be their the product space. If E and F in $\mathcal{X} \times \mathcal{Y}$ are such that $\nu(E_x) = \nu(F_x)$ for (almost) every $x \in X$, then show that $\pi(E) = \pi(F)$.

Hint: Apply the Tonelli Theorem to the characteristic functions χ_E and χ_F of the measurable sets E and F. Recall Problem 9.8.

Problem 9.15. Let $f, g \in \mathcal{L}(\mathbb{R}, \mathfrak{R}, \lambda)$ be Lebesgue integrable functions and consider the function $h: \mathbb{R}^2 \to \mathbb{R}$ given by $h(x,y) = f(x-y)$ for every (x,y)

in $\mathbb{R}^2$ that, according to Problem 9.2, is $\mathfrak{R}\times\mathfrak{R}$-measurable. Show that the function $hg\colon \mathbb{R}^2 \to \mathbb{R}$, mapping each $(x,y) \in \mathbb{R}^2$ into $f(x-y)g(y) \in \mathbb{R}$, also is $\mathfrak{R}\times\mathfrak{R}$-measurable. Now apply the Fubini Theorem to show that there is a real-valued function $f * g \in \mathcal{L}(\mathbb{R}, \mathfrak{R}, \lambda)$ defined almost everywhere on $\mathbb{R}$ by

$$(f * g)(x) = \int_{\mathbb{R}} h_x \, dy = \int_{\mathbb{R}} f(x-y)g(y) \, dy,$$

which is called the *convolution* of f and g. Moreover, also show that

$$\int |f * g| \, dx \le \left(\int |f| \, dx \right) \left(\int |g| \, dy \right).$$

Problem 9.16. Consider a countable family $\{\alpha_{m,n}\}$ of real numbers doubly indexed by $(m,n) \in \mathbb{N}\times\mathbb{N}$. Suppose $\alpha_{m,n} \ge 0$ for all $(m,n) \in \mathbb{N}\times\mathbb{N}$.

(a) Use elementary analysis to show that

$$\sum_{m=1}^{\infty} \sum_{n=1}^{\infty} \alpha_{m,n} = \sum_{n=1}^{\infty} \sum_{m=1}^{\infty} \alpha_{m,n}.$$

Hint: Recall that *a family of real numbers is absolutely summable if and only if it is unconditionally summable.*

Let $(\mathbb{N}, \wp(\mathbb{N}), \mu)$ be the measure space, where μ is the counting measure of Example 2B. Take the product space $(\mathbb{N}\times\mathbb{N}, \wp(\mathbb{N}\times\mathbb{N}), \mu\times\mu)$ of two copies of $(\mathbb{N}, \wp(\mathbb{N}), \mu)$. Consider the integral of $\alpha_{m,n}$ with respect to the product measure $\pi = \mu\times\mu$ (see Problem 3.4). That is, for every $E \in \wp(\mathbb{N}\times\mathbb{N})$,

$$\int_E \alpha_{m,n} \, d\pi = \sum_E \alpha_{m,n}.$$

(b) Show that if $\sum_{\mathbb{N}\times\mathbb{N}} \alpha_{m,n} < \infty$ (i.e., under the assumption of finite integral), the result in (a) can be proved by using the Fubini Theorem.

Problem 9.17. Consider the setup of the previous problem, but now dismiss the nonnegativeness assumption. Suppose $\alpha_{m,m} = 1$ and $\alpha_{m,m+1} = -1$ for every $n \in \mathbb{N}$, and $\alpha_{m,n} = 0$ otherwise. Show that

$$\sum_{m=1}^{\infty} \sum_{n=1}^{\infty} \alpha_{m,n} = 0 \quad \text{and} \quad \sum_{n=1}^{\infty} \sum_{m=1}^{\infty} \alpha_{m,n} = 1.$$

Thus conclude that the nonnegativeness assumption (i.e., $\alpha_{m,n} \ge 0$ for all (m,n) in $\mathbb{N}\times\mathbb{N}$) cannot the dropped in the Tonelli Theorem, and integrability (i.e., $\sum_{\mathbb{N}\times\mathbb{N}} |\alpha_{m,n}| < \infty$) cannot be omitted from the Fubini Theorem.

Hint:

$$\{\alpha_{m,n}\} = \begin{pmatrix} 1 & -1 & 0 & \\ 0 & 1 & -1 & 0 \\ & 0 & 1 & -1 \\ & & 0 & 1 \\ & & & & \ddots \end{pmatrix}.$$

Problem 9.18. Let $([0,1]\times[0,1], \wp([0,1]\times[0,1])\cap(\Re\times\Re), \lambda\times\lambda)$ be the product space obtained by two copies of the finite Lebesgue measure space $([0,1], \wp([0,1])\cap\Re, \lambda)$. Take $f \in \mathcal{L}([0,1], \wp([0,1])\cap\Re, \lambda)$ and the real-valued $g \in \mathcal{M}([0,1], \wp([0,1])\cap\Re)^+$ given by

$$f(x) = \begin{cases} 1, & 0 < x \le \frac{1}{2}, \\ -1, & \frac{1}{2} < x \le 1, \end{cases} \quad \text{and} \quad g(y) = \begin{cases} 0, & y = 0, \\ \frac{1}{y}, & y \in (0,1]. \end{cases}$$

Take the real-valued $h \in \mathcal{M}([0,1]\times[0,1], \wp([0,1]\times[0,1])\cap(\Re\times\Re))$ given by

$$h(x,y) = f(x)\,g(y)$$

for every $(x,y) \in [0,1]\times[0,1]$ (cf. Problem 9.3). Show that

$$\int_0^1 \left(\int_0^1 h(x,y)dx \right) dy = 0.$$

What is the value of the other iterated integral of h? Is h integrable?

Problem 9.19. An integrable function f (i.e., a function f in $\mathcal{L}(X,\mathcal{X},\mu)$) is a *real-valued* measurable function with a *finite integral* (i.e., $\int f^\pm d\mu < \infty$ or, equivalently, $\int |f| d\mu < \infty$ — see Lemma 4.4). Theorem 9.9 assumes that h is integrable and concludes that there exist integrable functions f_h and g_h such that $\int_X f_h d\mu = \int_{X\times Y} h d\pi = \int_Y g_h d\nu$, where μ and ν are σ-finite measures. As integrable functions, these h, f_h, and g_h are real-valued. Now show that, if μ and ν are σ-finite and if h is a measurable function (not necessarily real-valued) with a finite integral, then there exist measurable functions f_h and g_h (not necessarily real-valued but defined as in Theorem 9.9) with finite integrals such that $\int_X f_h d\mu = \int_{X\times Y} h d\pi = \int_Y g_h d\nu$. That is, the Fubini Theorem holds if we allow extended real-valued functions but retain the assumption of finite integrals (i.e., $\int_{X\times Y} |h| d\pi < \infty$).

Problem 9.20. Let $(X\times Y, \mathcal{X}\times\mathcal{Y}, \pi)$ be the product space of the measure spaces $(X,\mathcal{X},\mu)$ and $(Y,\mathcal{Y},\nu)$, where $\pi = \mu\times\nu$. Let f and g be real-valued functions on X and Y, respectively. Consider the real-valued function h on

$X \times Y$ defined by $h(x, y) = f(x) g(y)$ for every (x, y) in $X \times Y$. Show that if $f \in \mathcal{L}(X, \mathcal{X}, \mu)$ and $g \in \mathcal{L}(Y, \mathcal{Y}, \nu)$, then $h \in \mathcal{L}(X \times Y, \mathcal{X} \times \mathcal{Y}, \pi)$ and

$$\int_{X \times Y} h \, d\pi = \left(\int_X f \, d\mu \right) \left(\int_Y g \, d\nu \right)$$

even if the measures μ and ν are not σ-finite.

Hint: Put $F_0 = \{x \in X : f(x) = 0\}$ in $\mathcal{X}$ and $G_0 = \{y \in Y : g(y) = 0\}$ in $\mathcal{Y}$ so that $X' = X \backslash F_0 \in \mathcal{X}$ and $Y' = Y \backslash G_0 \in \mathcal{Y}$ are σ-finite sets with respect to μ and ν, respectively, according to Problem 3.9(c). Let $\mathcal{X}' = \wp(X') \cap \mathcal{X}$ and $\mathcal{Y}' = \wp(Y') \cap \mathcal{Y}$ be the sub-σ-algebras of $\mathcal{X}$ and $\mathcal{Y}$ consisting of subsets of X' and Y', and verify that μ and ν (actually, their restrictions to $\mathcal{X}'$ and $\mathcal{Y}'$) are σ-finite measures on $\mathcal{X}'$ and $\mathcal{Y}'$. Since $h = fg$, we get

$$\int_{X \times Y} h(x, y) \, d\pi = \int_{X' \times Y'} f(x) g(y) \, d\pi.$$

Thus we can apply the Fubini Theorem for σ-finite measures on $\mathcal{X}'$ and $\mathcal{Y}'$.

Suggested Reading

Bartle [3], Bauer [5], Berberian [6], Halmos [13], Royden [22].

References

1. S. ABBOTT, *Understanding Analysis*, Springer, New York, 2001.

2. E. ASPLUND AND L. BUNGART, *A First Course in Integration*, Holt, Rinehart and Winston, New York, 1966.

3. R.G. BARTLE, *The Elements of Integration*, Wiley, New York, 1966; enlarged 2nd ed., *The Elements of Integration and Lebesgue Measure*, Wiley, New York, 1995.

4. R.G. BARTLE, *The Elements of Real Analysis*, 2nd ed., Wiley, New York, 1976.

5. H. BAUER, *Measure and Integration Theory*, Walter de Gruyter, Berlin, 2001.

6. S.K. BERBERIAN, *Measure and Integration*, Chelsea, New York, 1965.

7. A. BROWN AND C. PEARCY, *Introduction to Operator Theory I: Elements of Functional Analysis*, Springer, New York, 1977.

8. A. BROWN AND C. PEARCY, *An Introduction to Analysis*, Springer, New York, 1995.

9. J. DUGUNDJI, *Topology*, Allyn & Bacon, Boston, 1960.

10. N. DUNFORD AND J.T. SCHWARTZ, *Linear Operators — Part I: General Theory*, Interscience, New York, 1958.

11. C. GOFFMAN AND G. PEDRICK, *A First Course in Functional Analysis*, 2nd ed., Chelsea, New York, 1983.

12. P.R. HALMOS, *Naive Set Theory*, Van Nostrand, New York, 1960; reprinted: Springer, New York, 1974.

13. P.R. HALMOS, *Measure Theory*, Van Nostrand, New York, 1950; reprinted: Springer, New York, 1974.

14. J.L. KELLEY, *General Topology*, Van Nostrand, New York, 1955; reprinted: Springer, New York, 1975.

15. J.L. KELLEY AND T.P. SRINIVASAN, *Measure and Integral — Volume 1*, Springer, New York, 1988.

16. A.N. KOLMOGOROV AND S.V. FOMIN, *Introductory Real Analysis*, Prentice-Hall, Englewood Cliffs, 1970.

17. C.S. KUBRUSLY, *Elements of Operator Theory*, Birkhäuser, Boston, 2001.

18. J.N. MCDONALD AND N.A. WEISS, *A Course in Real Analysis*, Academic Press, San Diego, 1999.

19. M.E. MUMROE, *Introduction to Measure and Integration*, Addison-Wesley, Reading, 1953.

20. M. REED AND B. SIMON, *Methods of Modern Mathematical Physics I: Functional Analysis*, 2nd ed., Academic Press, New York, 1980.

21. F. RIESZ AND B. SZ.-NAGY, *Functional Analysis*, Frederick Ungar, New York, 1955.

22. H.L. ROYDEN, *Real Analysis*, 3rd ed., Macmillan, New York, 1988.

23. W. RUDIN, *Real and Complex Analysis*, 3rd ed., McGraw-Hill, New York, 1987.

24. G.E. SHILOV AND B.L. GUREVICH, *Integral, Measure and Derivative: A Unified Approach*, Prentice-Hall, Englewood Cliffs, 1966.

25. P. SUPPES, *Axiomatic Set Theory*, Dover, New York, 1963.

26. R.L. VAUGHT, *Set Theory: An Introduction*, 2nd ed., Birkhäuser, Boston, 1995.

27. A.J. WEIR, *Lebesgue Integration and Measure*, Cambridge University Press, Cambridge, 1973.

28. A.J. WEIR, *General Integration and Measure*, Cambridge University Press, Cambridge, 1974.

Index

Printed and bound by CPI Group (UK) Ltd, Croydon, CR0 4YY

16/10/2024

01774904-0001